生态文明经典导读

张军亮　马　明　李湘慧　主编

南京大学出版社

图书在版编目(CIP)数据

生态文明经典导读：英文 / 张军亮，马明，李湘慧主编． -- 南京 ：南京大学出版社，2023.8
　ISBN 978-7-305-27252-3

Ⅰ．①生… Ⅱ．①张… ②马… ③李… Ⅲ．①生态环境建设－中国－英文 Ⅳ．①X321.2

中国国家版本馆CIP数据核字(2023)第160995号

出版发行　南京大学出版社
社　　址　南京市汉口路22号　邮　编　210093
书　　名　生态文明经典导读
　　　　　SHENGTAI WENMING JINGDIAN DAODU
主　　编　张军亮　马　明　李湘慧
责任编辑　孙　辉　　　　编辑热线 021-65248107

印　　刷　徐州绪权印刷有限公司
开　　本　889mm×1194mm　1/16　印张10.5　字数228千
版　　次　2023年8月第1版　2023年8月第1次印刷
ISBN 978-7-305-27252-3
定　　价　40.00元

网　　址　http://www.njupco.com
官方微博　http://weibo.com/njupco
官方微信号　njupress
南大悦学公众号　NJUyuexue
销售咨询热线　025-83594756

＊　版权所有，侵权必究
＊　凡购买南大版图书，如有印装质量问题，请与所购
　　图书销售部门联系调换

编委名单

副主编

郝丽丽　刘雪娇　马　亮　王　军　王　成
张维媛　范雅雯　郭乃源　闫光明　薛　扬

顾　问

李克国　崔力拓　韩宝军　王　璐　何　鑫

序言

生态文明，作为人类文明进步的重要形态，其核心理念在于实现人与自然、人与社会的和谐共生。在工业化浪潮的推动下，人类与自然之间的关系日趋紧张，生态环境问题日益严峻，成为全球可持续发展的关注焦点。因此，弘扬生态文明理念，培育生态文化，提高公众生态环境素养，让绿色低碳生活方式成风化俗，已成为全面推进美丽中国建设的必由之路。

生态兴的基石在于教育兴。加强环境教育，传播生态美学，在人们心中播撒绿色种子，从而激发全社会共同呵护生态环境的内生动力，尤为重要。作为一所以生态环境教育为办学特色的普通本科院校，河北环境工程学院传承和弘扬环保精神，不断探索和创新教育形式，将习近平生态文明思想有机融入教育教学全过程，浸润式培养学生生态文明意识。

本书作为学校校本课程双语读本，从环境保护的经典之作出发，引导学生深入理解和领会环境保护的丰富内涵，详细介绍了习近平生态文明思想的核心理念，并通过生态修复案例生动展示了建设美丽中国的显著成效，旨在激发青年学生对生态文明建设的热忱与动力。通过本书，引导学生树立并践行"绿水青山就是金山银山"的理念，成为生态文明建设的积极参与者和有力推动者，为广泛传播人与自然和谐共生的理念、全面推进美丽中国建设贡献智慧和力量。

李晓华

河北环境工程学院党委书记

Chapter I 第一章	Environment and Environmental Issues / 1 环境与环境问题 / 2
Section i 第一节	*Silent Spring* / 3 《寂静的春天》 / 7
Section ii 第二节	*Limits to Growth* / 10 《增长的极限》 / 14
Section iii 第三节	Declaration on the Human Environment / 16 《人类环境宣言》 / 20
Section iv 第四节	*Our Common Future* / 22 《我们共同的未来》 / 25
Section v 第五节	*Environmental Management in China* / 27 《中国环境管理》 / 32
Section vi 第六节	*China's Road of Green Development* / 35 《中国绿色发展之路》 / 39
Section vii 第七节	*Policies and Actions on Sustainable Development in China* / 41 《中国可持续发展政策与行动》 / 45
Section viii 第八节	*General Theory of Environmental Protection* / 47 《环境保护通论》 / 51
Section ix 第九节	On the Relation between Sustainable Development of Chinese Ancient Civilization and Ecologic Environment / 53 《论中国古代文明的可持续发展与生态环境的关系》 / 58
Section x 第十节	Sustainable Management of Water Resources / 61 《水资源的可持续管理》 / 65

Chapter II 第二章 — Xi Jinping Thought on Eco-civilization / 67
习近平生态文明思想 / 68

Section i 第一节
Harmony between Man and Nature / 69
人与自然和谐共生 / 72

Section ii 第二节
Lucid Waters and Lush Mountains Are Invaluable Assets / 74
绿水青山就是金山银山 / 80

Section iii 第三节
A Good Eco-environment Is the Most Inclusive Form of Public Wellbeing / 83
良好的生态环境是最普惠的民生福祉 / 87

Section iv 第四节
Mountains, Rivers, Forests, Farmlands, Lakes and Grasslands Are a Community of Life / 89
山水林田湖草是生命共同体 / 92

Section v 第五节
Protect the Ecological Environment with the Strictest System and the Strictest Rule of Law / 94
用最严格制度、最严密法治保护生态环境 / 97

Section vi 第六节
Work Together to Promote a Global Eco-civilization / 99
共谋全球生态文明建设 / 103

Chapter III 第三章 — Amazing China / 105
美丽中国 / 106

Section i 第一节
Saihanba / 107
塞罕坝 / 111

Section ii 第二节
Yucun / 113
余村 / 118

Section iii 第三节
Mangrove to Gold Forest / 121
红树林是金树林 / 124

Section iv 第四节
Water, Soil Conservation—Changting, Fujian / 126
中国水土流失综合治理——福建长汀 / 131

Section v 第五节
Yunnan Snub-nosed Monkey Range-wide Conservation / 134
滇金丝猴全境保护 / 138

Section vi 第六节
Ecological Restoration of Qingxi Country Park in Shanghai / 140
上海青西郊野公园生态修复 / 145

Section vii 第七节
Carbon Peak and Carbon Neutrality / 148
碳达峰和碳中和 / 152

Section viii 第八节
China Embraces Garbage Classification, with Shanghai Taking the Lead / 154
中国采用垃圾分类——上海试点 / 158

Chapter I

Environment and Environmental Issues

Chapter I is divided into 10 sections, including 4 classic works on environmental protection in the world and 6 works of outstanding contributors to environmental protection in China selected in chronological order. From *Silent Spring*, a powerful warning, to *Limits to Growth*, a renewed call for timely action, from the Declaration on the Human Environment in 1972, to the signing of the implementing regulations of the Paris Agreement in 2021, the road of environmental protection has been moving forward in exploration. In 1973, China held the first environmental protection conference, which is a crucial step in the cause of environmental protection in China. Qu Geping, Xie Zhenhua and others with lofty ideals have made great contributions to promoting the development of environmental protection and education in China. Nowadays, China is playing a leading role in achieving green transformation and sustainable development around the world.

Through reading, reflection and activities, learners will appreciate ecological literature works, understand the core concepts of environmental protection, realize the significance of environmental protection, cultivate and practice the thought of eco-civilization.

第一章

环境与环境问题

第一章分为十小节,按时间顺序选取四本世界环境保护经典著作和六位中国环境保护事业杰出贡献者的论著。从《寂静的春天》给予人类强有力的警示,到《增长的极限》再一次呼吁我们及时采取行动,从1972年的《人类环境宣言》到2021年《巴黎协定》实施细则的签订,环境保护之路一直在探索中前行。1973年,我国举办了第一次环境保护会议,迈出了中国环境保护事业关键性的一步。曲格平、解振华等仁人志士致力于推动中国环保事业和环保教育的发展,并为此做出巨大的贡献。中国正在为全球实现绿色转型和可持续发展发挥引领作用。

通过本章文本细读、思考和拓展练习,学生将赏析生态文学作品,理解环境保护的核心概念,了解环境保护的举措和重要意义,培养和践行生态文明思想。

Section i

Silent Spring

Introduction

Published in 1962, *Silent Spring* is one of the most classic books on the topic of environmental protection. It actually stimulated and prompted the awakening of environmental protection awareness. Since then, a new age has started.

Silent Spring began with a "fable for tomorrow"—a true story using a composite of examples drawn from many real communities where the use of DDT (dichloro-diphenyl-trichloroethane) had caused damage to wildlife, birds, bees, agricultural animals, domestic pets, and even humans. Carson used it as an introduction to a very scientifically complicated and already controversial subject. This "fable" made an indelible impression on readers and was used by critics to charge that Carson was a fiction writer and not a scientist.

Silent Spring suggested a needed change in how democracies and liberal societies operated so that individuals and groups could question what their governments allowed others to put into the environment. Far from calling for sweeping changes in government policy, Carson believed the federal government was part of the problem. She admonished her readers and audiences to ask "Who speaks, and why?" and therein to set the seeds of social revolution. She identified human hubris and financial self-interest as the crux of the problem and asked if we could master ourselves and our appetites to live as though we humans are an equal part of the earth's systems and not the master of them.

Silent Spring inspired the modern environmental movement, which began in earnest a decade later. It is recognized as the environmental text that "changed the world". She aimed at igniting a democratic activist movement that would not only question the direction of science and technology but would also demand answers and accountability. Rachel Carson was a prophetic voice and her "witness for nature" is even more relevant and needed if our planet is to survive into a 22nd century.

Excerpts

A Fable for Tomorrow

There was once a town in the heart of America where all life seemed to live in harmony with its surroundings. The town lay in the midst of a checkerboard of prosperous farms, with fields of grain and hillsides of orchards where, in spring, white clouds of bloom drifted above the green fields. In autumn, oak and maple and birch set up a blaze of color that flamed and flickered across a backdrop of pines. Then foxes barked in the hills and deer silently crossed the fields, half hidden in the mists of the autumn mornings.

Chapter I 第一章

Along the roads, laurel, viburnum and alder, great ferns and wildflowers delighted the traveller's eye through much of the year. Even in winter the roadsides were places of beauty, where countless birds came to feed on the berries and on the seed heads of the dried weeds rising above the snow. The countryside was, in fact, famous for the abundance and variety of its bird life, and when the food of migrants was pouring through in spring and autumn people travelled from great distances to observe them. Others came to fish the streams, which flowed clear and cold out of the hills and contained shady pools where trout lay. So it had been from the days many years ago when the first settlers raised their houses, sank their wells, and built their barns.

Then a strange blight crept over the area and everything began to change. Some evil spell had settled on the community: mysterious maladies swept the flocks of chickens; the cattle and sheep sickened and died. Everywhere was a shadow of death. The farmers spoke of much illness among their families. In the town the doctors had become more and more puzzled by new kinds of sickness appearing among their patients. There had been several sudden and unexplained deaths, not only among adults but even among children, who would be stricken suddenly while at play and die within a few hours.

There was a strange stillness. The birds, for example—where had they gone? Many people spoke of them, puzzled and disturbed. The feeding stations in the backyards were deserted. The few birds seen anywhere were moribund; they trembled violently and could not fly. It was a spring without voices. On the mornings that had once throbbed with the dawn chorus of robins, catbirds, doves, jays, wrens, and scores of other bird voices there was now no sound; only silence lay over the fields and woods and marsh.

On the farms the hens brooded, but no chicks batched. The farmers complained that they were unable to raise any pigs—the litters were small and the young survived only a few days. The apple trees were coming into bloom but no bees droned among the blossoms, so there was no pollination and there would be no fruit.

The roadsides, once so attractive, were now lined with browned and withered vegetation as though swept by fire. These, too, were silent, deserted by all living things. Even the steams were now lifeless. Anglers no longer visited them, for all the fish bad died.

In the gutters under the eaves and between the shingles of the roofs, a white granular powder still showed a few patches; some weeks before it had fallen like snow upon the roofs and the lawns, the fields and streams.

No witchcraft, no enemy action had silenced the rebirth of new life in this stricken world. The people had done it themselves.

This town does not actually exist, but it might easily have a thousand counterparts in America or elsewhere in the world. I know of no community that has experienced all the misfortunes I describe. Yet every one of these disasters has actually happened somewhere, and many real communities have already suffered a substantial number of them.

A grim spectre has crept upon us almost unnoticed, and this imagined tragedy may easily become a stark reality we all shall know.

What has already silenced the voices of spring in countless towns in America? This book is an attempt to explain.

(Excerpted from Chapter One)

Vocabulary

abundance 大量；丰盛
robin 知更鸟
pollination 传粉

blight 枯萎病
survive 幸存；生存
witchcraft 巫术

Reflection and Activities

1. Although the incident of "the fable" did not actually happen in the modern world, there is this concern about the harm of chemical pollutions all the time, especially in today's China. Think about the world described in "the fable", what if it had happened, what would this world be like today and what would be China like today? Support your point with examples, statistics, etc.

2. The last chapter of *Silent Spring* is "The other road" which describes a later called "sustainable development" world. Consult the internet or your library to find out what is "the other road"? What does it have to do with our life today?

About the Author

Rachel Louise Carson (1907–1964) was born in America who is a marine biologist, writer, and conservationist. Her *Silent Spring* and other writings are so influential that the global environmental movements were advanced. As one of the greatest nature writer of the 20th century, Rachel Carson challenged the convention that humans could gain dominance over nature by chemicals, bombs and space travel. The book *Silent Spring* (1962) warned of the dangers to all natural ecosystems from the ill-use of chemical pesticides such as DDT, and questioned the range and direction of modern science, stimulated the contemporary environmental movement.

《寂静的春天》

导读

1962年出版的《寂静的春天》是关于环境保护的最经典的书籍之一。人们的环保意识逐渐觉醒。从此，一个新的时代开始了。

《寂静的春天》以一个"明天的寓言"开始——这是一个真实的故事，综合了许多真实社区的例子，在这些社区中，DDT（双对氯苯基三氯乙烷）的使用对野生动物、鸟类、蜜蜂、农业动物、家养宠物甚至人类造成了损害。这则"寓言"给读者留下了不可磨灭的印象，有评论家斥责卡森是一名小说家而不是科学家。

《寂静的春天》提出了民主国家和自由社会运作方式的必要变革，以便个人和团体可以质疑他们的政府允许民众将什么投入到环境中。卡森认为联邦政府不仅要改变政策，同时，联邦政府本身也是问题的一部分。她告诫读者，要问"谁在讲话，为什么？"并在其中播下社会革命的种子。她认为人类的自大和经济上的自利是问题的症结所在，并质疑我们是否能够控制自己和我们的欲望，让我们人类的生活成为地球系统的平等组成部分，而不是它的主宰。

《寂静的春天》点燃了现代环保运动的火花，而环保运动十年后才真正开始。它被认为是"改变了世界"的环保书籍。卡森旨在引发一场环境保护运动，这场运动不仅对科学和技术的方向提出质疑，还会要求解决办法和问责。雷切尔·卡森发出预言的呼声和大自然的见证，对于未来22世纪的生存空间是至关重要的。

节选

明天的预言

从前，在美国中部有一个城镇，这里的生物看起来与周围环境很和谐。这个城镇坐落在像棋盘般排列整齐的繁荣的农场中央，其周围是庄稼地，小山下果树成林。春天，繁花像白色的云朵点缀在绿色的原野上；秋天，透过松林的屏风，橡树、枫树和白桦闪射出火焰般的彩色光辉，狐狸在小山上叫着，小鹿静悄悄地穿过了笼罩着秋天晨雾的原野。

沿着小路生长的月桂树、荚蒾和赤杨树、以及巨大的羊齿植物和野花在一年的大部分时间里都使旅行者感到目悦神怡。即使在冬天，道路两旁也是美丽的地方，那儿有无数小鸟飞来，在出露于雪层之上的浆果和干草的穗头上啄食。郊外事实上正以其鸟类的丰富多样而驰名，当迁徙的候鸟在整个春天和秋天蜂拥而至的时候，人们都长途跋涉地来这里观看它们。另有些人来小溪边捕鱼，这些洁净又清凉的小溪从山中流出，形成了绿荫掩映的生活着鳟鱼的池塘。野外一直是这个样子，直到许多年前的有一天，第一批居民来到这儿建房舍、挖井筑仓，情况才发生了变化。

从那时起，一个奇怪的阴影遮盖了这个地区，一切都开始变化。一些不祥的预兆降临到村落里：神秘莫测的疾病袭击了成群的小鸡；大量牛羊病倒和死亡。到处是死神的幽灵。农夫们述说着他们家庭的多病。城里的医生也愈来愈为他们病人中出现的新病感到困惑莫解。不仅在成人中，而且在孩子中出现了一些突然的、不可解释的死亡现象，这些孩子在玩耍时突然倒下了，并在几小时内死去。

一种奇怪的寂静笼罩了这个地方。比如说，鸟儿都到哪儿去了呢？许多人谈论着它们，感到迷惑和不安。园后鸟儿寻食的地方冷落了。在一些地方仅能见到的几只鸟儿也气息奄奄，它们浑身颤抖，飞不起来。这是一个没有声息的春天。这儿的清晨曾经萦绕着知更鸟、猫鹊、鸽子、樫鸟、鹪鹩的合唱以及其他鸟鸣的音浪；而现在一切声音都没有了，只有一片寂静覆盖着田野、树林和沼地。

农场里，母鸡在孵窝，但却没有小鸡破壳而出。农夫们抱怨着他们无法再养猪了——新生的猪仔很小，小猪病后也只能活几天。苹果树花要开了，但在花丛中没有蜜蜂嗡嗡飞来，所以苹果花没有得到授粉，也不会有果实。

曾经一度是多么迷人的小路两旁，现在排列着仿佛火灾劫后的、焦黄的、枯萎的植物。被生命抛弃了的这些地方也是寂静一片。甚至小溪也失去了生命；钓鱼的人不再造访，因为所有的鱼已死亡。

在屋沿下的雨水管中，在房顶的瓦片之间，一种白色的粉粒还在露出稍许斑痕。在几星期之前，这些白色粉粒好像雪花一样降落到屋顶、草坪、田地和小河上。

不是魔法，也不是敌人的活动使这个受伤的世界的生命无法复生，而是人们害了自己。

上述的这个城镇是虚设的，但在美国和世界其他地方都可以轻易找到上千个这种城镇的翻版。我知道并没有一个村庄经受过如我所描述的全部灾祸；但其中每一种灾难实际上已在某些地方发生，并且确实有许多村庄已经承受了大量的不幸。

在人们的忽视中，一个狰狞的幽灵已向我们袭来，这个想象中的悲剧可能会变成一个活生生的现实。

是什么东西使得美国无数城镇的春天之音沉寂下来了呢？这本书试着给予解答。

（节选自第一章）

思考题

1.《寓言》事件虽然在现代社会并没有发生，但人们一直担心化学污染的危害。想想《寓言》

中描述的世界，如果它发生了，这个世界今天会是什么样子？中国今天会是什么样子？用例子、统计数据等来支持你的观点。

2.《寂静的春天》的最后一章是"另一条路"，它描述了一个后来被称为"可持续发展"的世界。查阅互联网或图书，了解"另一条路"是什么？这和我们今天的生活有什么关系？

关于作者

雷切尔·路易斯·卡森（1907—1964）出生于美国，是海洋生物学家、作家、自然保护主义者，她的著作《寂静的春天》等影响深远，推动了全球环保运动的发展。作为二十世纪最伟大的自然作家之一，雷切尔·路易斯·卡森对人类可以通过化学物质、炸弹和太空旅行统治自然的想法提出了挑战。雷切尔·路易斯·卡森在《寂静的春天》（1962）一书中警示了DDT等化学杀虫剂的不当使用对所有自然生态系统造成的危害，并对现代科学的范围和方向提出了质疑，引发了当代环保运动。

Limits to Growth

Introduction

In 1968, Dr. Aurelio Percy, chairman of the European Italian consulting company, invited 30 scientists, educators, economists and politicians from more than 10 developed countries to form an informal international association Club of Rome at the Lindsey Institute in Rome to study the current and future situation of mankind. Entrusted by the Rome club, the research team led by MIT scholar Dennis Meadows conducted an empirical study on five variables: population, agricultural production, natural resources, industrial production and pollution based on computer model and system dynamics, and submitted its first report, *The Limits to Growth*, in 1972. The report was published as a book and became one of the largest books that changed the world in the 20th century, which also contributed to the birth and development of the concept of "sustainable development" later.

You can believe nothing, but you must believe the facts. 30 years ago, when people criticized the *The Limits to Growth*, 30 years later, you can no longer believe the fact that the limit of growth, because a series of problems such as environmental pollution and resource shortage caused by the limit of growth have appeared in front of us.

Looking back on history, in the era of industrialization, Western countries blindly pursue the growth of interests and ignore environmental benefits. Although they have reached the "golden age" of the west, this golden age is short, and then there is a crisis. We have seen that behind the high prosperity of Western countries, various environmental problems are becoming increasingly apparent.

Since the reform and opening up, China has always adhered to economic development as the first priority, ignoring the protection of the environment and a large amount of resource consumption. Only from the current rise in oil prices, we can see that the serious resource bottleneck has made China face the limit of economic development. It is estimated that in the next few decades, China will become the world's largest resource consumer. If effective measures are not taken to protect resources, what will be the guarantee for our future development?

"Hardship endangers life and complacency brings demise" is a truth that has been preached for thousands of years. Everyone knows it, but everyone always ignores this truth when doing things. *The Limits to Growth* once again sounded the alarm to us, calling on us to take timely action before the earth's limit comes.

Therefore, in order to avoid repeating the limit that Western developed countries pay one-sided attention to economic development, we must protect the environment and take the road of sustainable development while developing the economy. That is, while achieving the purpose of economic development, we should protect the natural resources and environment on which good people depend, so as not only to meet the needs of our contemporary people, but also not to endanger the needs of future generations.

In addition, the environmental problem is no longer a local problem. On the contrary, it is showing a trend of global diffusion. It can not be solved by relying on the efforts of each place. It requires the joint efforts of people all over the world. Therefore, it is necessary to strengthen the cooperation between various countries to jointly deal with and solve the environmental crisis.

Excerpts

Overshoot

To overshoot means to go too far, to go beyond limits accidentally—without intention. People experience overshoots every day. When you rise too quickly from a chair, you may momentarily lose your balance. If you turn on the hot-water faucet too far in the shower, you may be scalded. On an icy road your car might slide past a stop sign. At a party you may drink much more alcohol than your body can safely metabolize; in the morning you will have a ferocious headache. Construction companies periodically build more condominiums than are demanded, forcing them to sell units below cost and confront the possibility of bankruptcy. Too many fishing boats are often constructed. Then fishing fleets grow so large that they catch for more than the sustainable harvest. This depletes the fish population and forces ships to remain in harbor. Chemical companies have produced more chlorinated chemicals than the upper atmosphere can safely assimilate. Now the ozone layer will be dangerously depleted for decades until stratospheric chlorine levels decline.

(Excerpted from Chapter One)

The Driving Force: Exponential Growth

I find to my personal horror that I have not been immune to naivete about exponential functions... While I have been aware that the interlinked problems of loss biological diversity, tropical deforestation, forest die back in the northern hemisphere and climate change are growing exponentially, it is only this very year that I think I have truly internalized how rapid their accelerating threat really is.

(Excerpted from Chapter Two)

The Limits: Sources and Sinks

The technologies we adopted that enabled us to maintain constant or declining dollar costs for resources often required ever-increasing amounts of direct and indirect fuel...this luxury becomes costly necessity, requiring that increasing proportions of our national income be diverted to the resource-processing sectors in order to supply the same quantity of resource.

(Excerpted from Chapter Three)

World3: The Dynamics of Growth in a Finite World

If current prediction growth prove accurate and patterns of human activity on the planet remain unchanged, science and technology may not be able to prevent either irreversible degradation of the environment or continued poverty for much of the world.

(Excerpted from Chapter Four)

Vocabulary

overshoot 过冲

oscillation 震荡

exponential growth 指数型增长

ecological footprint 生态足迹

delay 时滞

collapse 崩溃

Reflection and Activities

1. After class debate: Should human beings conform to nature, reduce material desires, be Buddha-like and lying flat; Or actively create, innovate in science and technology, find new growth points, develop the economy rapidly, and let the people live a better material and cultural life?

2. Under what domestic and international background does China propose the dual carbon goal and what is the timeline? Please draw a mind map.

第二节 《增长的极限》

导读

1968年，欧洲-意大利咨询公司主席奥雷里奥·珀西博士邀请来自10多个发达国家的30名科学家、教育家、经济学家和政界人士在罗马林赛研究所成立一个非正式的国际协会罗马俱乐部，研究人类的现状和未来状况。受罗马俱乐部的委托，麻省理工学院学者丹尼斯·梅多斯领导的研究团队基于计算机模型和系统动力学，对人口、农业生产、自然资源、工业生产和污染五个变量进行了实证研究，并于1972年提交了第一份报告《增长的极限》。该报告以书的形式出版，成为20世纪改变世界的最大书籍之一，这也促进了"可持续发展"概念的诞生和发展。

你可以什么都不信，但一定要相信事实。三十年前，人们诟病《增长的极限》；三十年后，因为增长的极限所带来的一系列诸如环境污染、资源短缺等问题已经呈现在我们面前。

当我们回顾历史时不难发现，西方国家在工业化时代中盲目追求经济增长而无视环境利益。尽管西方曾经出现过"流金岁月"，但它十分短暂，随之而来的则是一场危机。我们见证了西方国家高度繁荣的背后，各种环保问题层出不穷。

中国在改革开放后持续将经济发展作为第一要务，忽视了对环境的保护，也造成了大量能源的消耗。直到最近的原油价格上涨才让我们意识到严重的资源瓶颈已经使中国面临经济发展的极限。估计在未来的几十年里，中国将成为世界上最大的能源消费国。如果不采取有效的手段保护资源，那还如何能保证未来的发展？

"生于忧患而死于安乐"是我们相传数千年的真理。人人都知道这个真理，但人人在做事时都选择了忽视。《增长的极限》再一次为我们长鸣警钟，呼吁我们在地球的极限到来前及时采取行动。

因此，为了避免重蹈西方发达国家一边倒发展经济的覆辙，我们要在发展经济的同时保护环境，走可持续发展的道路。这就是说，一方面要达到经济发展的目的，一方面要依靠人民保护自然资源和环境，不仅要满足当代人民群众的需求，还不能危及子孙后代。

除此之外，环境问题也不再是一个区域的问题。恰恰相反，它呈现了一种全球蔓延的态势。环境问题的解决不能只靠一时一地的努力，它需要全球人民的合力。因此，有必要加强各个国家的合作来共同应对和解决环境危机。

Chapter I 第一章

节选

过 冲

过冲的意思就是走的太远，偶然超越了极限却又毫无意识。人们每天都能经历过冲。当你从椅子上起的太急，很可能就在一瞬间失去平衡。如果你在淋浴的时候一下把花洒开得太大就有可能被烫伤。在结冰的路面你的车很可能遇到停止标线也刹不住。在聚会上你可能因为饮酒过量超过自己新陈代谢的限值，早上起床就会头痛欲裂。建筑公司总是周期性地建起过量的公寓，最终被迫以低于成本的价格卖出，或者面临破产的风险。我们也造了过量的渔船，然后又会因为渔船越造越大不得不去大肆捕捞而终将无法持续。最终渔业资源耗尽迫使渔船不再离港。化工企业总是过量生产高层大气无法安全吸收的氯化物。未来的几十年臭氧层还会以非常可怕的速度减少，直到平流层的氯含量降低为止。

（节选自第一章）

驱动力：指数增长

我震惊地发现我对指数增长仍然抱有天真的想法……然而直到我意识到植物多样性的丧失、热带雨林的滥砍滥伐、北半球的森林退化和气候变化的指数型增长，我才在今年真正在内心深处认识到了真正的威胁进展得有多么迅速。

（节选自第二章）

极限：源与汇

我们所采用的能让我们保持平稳或降低资源成本消耗的技术经常需要不断增加直接和间接的能源量……这种奢侈已经成为必要消费，这就要求我们增加国家收入转化为资源加工部门的比例，而这样做就只是为了提供相同数量的资源。

（节选自第三章）

World3 模型：有限世界里的增长动力

如果当前的增长预测被证明准确，并且地球上人类活动的模式保持不变，科学和技术可能无法阻止环境不可逆转的恶化，也无法阻止世界大部分地区持续贫困。

（节选自第四章）

思考题

1. 课后讨论：人类到底应该顺应自然、降低物欲、佛系、躺平，还是要积极创造、科技创新寻找新的增长点，快速发展经济，让人民过上更加丰富的物质文化生活？

2. 中国给出双碳目标承诺的国内和国际背景是什么？时间线是什么？请画出实现双碳目标时间线的思维导图。

Declaration on the Human Environment

Introduction

On June 5, 1972, the United Nations held the first Human Environment Conference in Stockholm, Sweden. It passed the famous "Declaration on the Human Environment" and the "Action Plan" for the protection of the global environment, expounded seven common views and 26 principles achieved by participating countries and international organizations and proposed the slogan "to protect this and future generations and improve the environment". In the same year, the United Nations General Assembly passed a resolution to establish June 5 every year as "World Environment Day". 133 countries in the world sent representatives to the United Nations Conference on the human environment in 1972. Chen Haifeng and Qu Geping attended the meeting on behalf of China.

In 1973, China held the first environmental protection conference. It has taken a key step in China's environmental protection. In 1983, China established environmental protection as a basic national policy, and in 2017, "Establishing and practicing the concept that lucid waters and lush mountains are invaluable assets" was written into the report of the 19th National Congress. Environmental protection has become a national strategy.

Excerpts

Declaration of the United Nations Conference on the Human Environment

The United Nations Conference on the Human Environment, having met at Stockholm from 5 to 16 June 1972, having considered the need for a common outlook and for common principles to inspire and guide the peoples of the world in the preservation and enhancement of the human environment, proclaims that:

1. Man is both creature and moulder of his environment, which gives him physical sustenance and affords him the opportunity for intellectual, moral, social and spiritual growth. In the long and tortuous

evolution of the human race on this planet a stage has been reached when, through the rapid acceleration of science and technology, man has acquired the power to transform his environment in countless ways and on an unprecedented scale. Both aspects of man's environment, the natural and the man-made, are essential to his wellbeing and to the enjoyment of basic human rights the right to life itself.

2. The protection and improvement of the human environment is a major issue which affects the wellbeing of peoples and economic development throughout the world; it is the urgent desire of the peoples of the whole world and the duty of all Governments.

3. Man has constantly to sum up experience and go on discovering, inventing, creating and advancing. In our time, man's capability to transform his surroundings, if used wisely, can bring to all peoples the benefits of development and the opportunity to enhance the quality of life. Wrongly or heedlessly applied, the same power can do incalculable harm to human beings and the human environment. We see around us growing evidence of man-made harm in many regions of the earth: dangerous levels of pollution in water, air, earth and living beings; major and undesirable disturbances to the ecological balance of the biosphere; destruction and depletion of irreplaceable resources; and gross deficiencies, harmful to the physical, mental and social health of man, in the man-made environment, particularly in the living and working environment.

4. In the developing countries most of the environmental problems are caused by under-development. Millions continue to live far below the minimum levels required for a decent human existence, deprived of adequate food and clothing, shelter and education, health and sanitation. Therefore, the developing countries must direct their efforts to development, bearing in mind their priorities and the need to safeguard and improve the environment. For the same purpose, the industrialized countries should make efforts to reduce the gap themselves and the developing countries. In the industrialized countries, environmental problems are generally related to industrialization and technological development.

5. The natural growth of population continuously presents problems for the preservation of the environment, and adequate policies and measures should be adopted, as appropriate, to face these problems. Of all things in the world, people are the most precious. It is the people that propel social progress, create social wealth, develop science and technology and, through their hard work, continuously transform the human environment. Along with social progress and the advance of production, science and technology, the capability of man to improve the environment increases with each passing day.

6. A point has been reached in history when we must shape our actions throughout the world with a more prudent care for their environmental consequences. Through ignorance or indifference we can

do massive and irreversible harm to the earthly environment on which our life and wellbeing depend. Conversely, through fuller knowledge and wiser action, we can achieve for ourselves and our posterity a better life in an environment more in keeping with human needs and hopes. There are broad vistas for the enhancement of environmental quality and the creation of a good life. What is needed is an enthusiastic but calm state of mind and intense but orderly work. For the purpose of attaining freedom in the world of nature, man must use knowledge to build, in collaboration with nature, a better environment. To defend and improve the human environment for present and future generations has become an imperative goal for mankind—a goal to be pursued together with, and in harmony with, the established and fundamental goals of peace and of worldwide economic and social development.

7. To achieve this environmental goal will demand the acceptance of responsibility by citizens and communities and by enterprises and institutions at every level, all sharing equitably in common efforts. Individuals in all walks of life as well as organizations in many fields, by their values and the sum of their actions, will shape the world environment of the future.

Local and national governments will bear the greatest burden for large-scale environmental policy and action within their jurisdictions. International cooperation is also needed in order to raise resources to support the developing countries in carrying out their responsibilities in this field. A growing class of environmental problems, because they are regional or global in extent or because they affect the common international realm, will require extensive cooperation among nations and action by international organizations in the common interest.

The Conference calls upon Governments and peoples to exert common efforts for the preservation and improvement of the human environment, for the benefit of all the people and for their posterity.

(Excerpted from seven common views expounded on the conference)

Vocabulary

The First Human Environment Conference 第一次人类环境会议
Declaration on the Human Environment 人类环境宣言
World Environment Day 世界环境日
International Mother Earth Day 国际地球母亲日

Reflection and Activities

1. Read United Nations Secretary General Antonio Guterres's excerpt from the 2020 International Earth Day, then make a speech in this year's International Earth Day, assuming you are the head of the school environmental protection association.

2. In the National Conference on ecological and environmental protection in 2022, the environmental protection policy of "precise pollution control, scientific pollution control and pollution control according to law" was emphasized. What do you think can college students do for environmental proection?

《人类环境宣言》

导读

1972年6月5日,联合国在瑞典首都斯德哥尔摩举行第一次人类环境会议,通过了著名的《人类环境宣言》及保护全球环境的"行动计划",阐明了与会国和国际组织所取得的七个共识和二十六项原则,提出"为了子孙后代,保护环境"的口号。同年,联合国大会通过决议,将每年的6月5日设立为"世界环境日"。133个国家派代表参加了此次联合国人类环境会议。曲格平、陈海峰等代表中国参加了此次会议。

1973年,我国举办了第一次环境保护会议,迈出了中国环境保护事业关键性的一步。我国在1983年将保护环境立为基本国策,2017年又将"树立和践行绿水青山就是金山银山的理念"写进十九大报告,环境保护成为国家战略。

节选

联合国人类环境会议宣言

联合国人类环境大会于1972年6月5日至16日在斯德哥尔摩举行会议,审议了一项共同的展望和原则,以激励和引导世界各国人民保护、改善人类环境,兹宣布:

1. 人是环境的产物,也是环境的塑造者。由于当代科学技术突飞猛进的发展,人类已获得了无数方法和在空前的规模上改造环境的能力。自然环境和人为环境对于人的福祉和基本人权,都是必不可少的。

2. 保护和改善人类环境关系到各国人民的福利和经济发展,是人民的迫切愿望,是各国政府应尽的责任。

3. 人类总是要不断地总结经验,有所发现,有所发明,有所创造,有所前进。人类改变环境的能力,如妥善地加以运用,可为人类带来福祉;如运用不当,则对人类和环境造成不可估量的损害。现在地球上许多地区出现日益加剧危害环境的迹象,在人为环境,特别是生活和工作环境中也已经出现了有害人体身心健康的重大缺陷。

4. 在发展中国家,多数的环境问题是发展不足引起的。因此,它们首先要致力于发展,同时也要顾及保护和改善环境。在工业发达国家，环境问题一般是由工业和技术发展产生的。

5. 人口的自然增长不断引起环境问题，因此要采取适当的方针和措施，解决这些问题。

6. 当今的历史阶段要求世界上人们在计划行动时更加谨慎地考虑将给环境带来的后果。为了在自然界获得自由，人类必须运用知识，同自然取得协调，以便建设更良好的环境。为当代和子孙后代保护好环境已成为人类的迫切目标。这同和平、经济和社会的发展目标完全一致。

7. 为实现这个环境目标，要求每个公民、团体、机关、企业都负起责任，共同创造未来的世界环境。各国中央和地方政府对大规模的环境政策和行动负有特别重大的责任。对于区域性和全球性的环境问题，在共同利益的前提下，由各国进行广泛合作，由国际组织采取行动。

地方和国家政府将为其管辖范围内的大规模环境政策和行动承担最大的负担。还需要开展国际合作，以筹集资源，支持发展中国家履行其在这一领域的责任。不断增多的环境问题需要各国以及国际组织的通力合作，为了共同利益采取行动，因为如今的环境问题大多是区域性或全球性的，或是影响着国际公共领域。

会议呼吁各国政府和人民为保护和改善人类环境作出共同努力，造福于全体人民和子孙后代。

（节选自对会议阐述的七个共识）

思考题

1. 阅读联合国秘书长安东尼奥·古特雷斯在 2020 年"国际地球母亲日"的发言节选，思考：假设你是学校环境保护协会的负责人，你如何在今年的"国际地球母亲日"活动中发言？

2. 2022 年的全国生态环境保护工作会议，重点强调了"精准治污、科学治污、依法治污为工作方针"的环保方针。结合所学专业，谈一谈在校大学生能够为保护环境做什么？

Section iv *Our Common Future*

Introduction

Do you know the concept of "sustainable development"?

—Humanity has the ability to make development sustainable—to ensure that it meets the needs of the present without compromising the ability of future generations to meet their own needs.

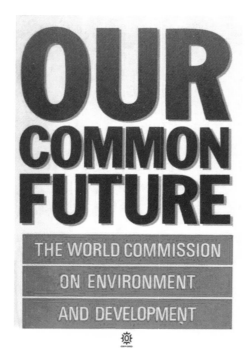

Our Common Future is the report of the World Commission on environment and development about the future of mankind. It was approved at the 8th World Commission on Environment and Development held in Tokyo, Japan in February 1987. After a debate in General Assembly of the United Nations in April, 1987, *Our Common Future* was officially published in April 1987.

This key concept of sustainable development was given in Gro Brundtland's *Our Common Future* (1987) for the first time. The publication of the book marked an important beginning in our history.

The report has greatly changed international political thought and prompted people to recognize our responsibility, use natural resources responsibly and benefit future generations.

The report received much positive attention immediately after publication and was hailed by Oxford University Press as "the most important document of the decade on the future of the world". Indeed, it is possible to expand this point by claiming the report to be one of the most influential texts of the twentieth century.

Under the leadership of Mrs. Brundtland, 21 members of the World Commission on environment and development systematically studied the major economic, social and environmental problems faced by mankind. Taking "sustainable development" as the guiding principle, they put forward a series of policy objectives and action suggestions from the perspective of protecting and developing environmental resources and meeting the needs of present and future generations.

The report combines environment and development as a whole. The guiding ideology set forth in the report is a positive and significant reference when governments and people of all countries formulate a policy.

Our Common Future points out that governments and people all over the world must take their historical responsibility for the two major issues of economic development and environmental protection from now on, formulating and implementing the correct policies. Serious damage to the ecological environment has occurred, and immediate action must be taken.

The report provides the first comprehensive overview of major issues such as rapid world population growth, food shortage, urban population growth, environmental degradation and climate change. The report affects today's policy-making process. The report's reference to meeting major challenges, such as climate change, remains relevant today.

Excerpts

Sustainable Development

Humanity has the ability to make development sustainable—to ensure that it meets the needs of the present without compromising the ability of future generations to meet their own needs. The concept of sustainable development does imply limits—not absolute limits but limitations imposed by the present state of technology and social organization on environmental resources and by the ability of the biosphere to absorb the effects of human activities. But technology and social organization can be both managed and improved to make way for a new era of economic growth. The Commission believes that widespread poverty is no longer inevitable. Poverty is not only an evil in itself, but sustainable development requires meeting the basic needs of all and extending to all the opportunity to fulfil their aspirations for a better life. A world in which poverty is endemic will always be prone to ecological and other catastrophes.

Meeting essential needs requires not only a new era of economic growth for nations in which the majority are poor, but an assurance that those poor get their fair share of the resources required to sustain that growth. Such equity would be aided by political systems that secure effective citizen participation in decision making and by greater democracy in international decision making.

Sustainable global development requires that those who are more affluent adopt lifestyles within the planet's ecological means—in their use of energy, for example. Further, rapidly growing populations can increase the pressure on resources and slow any rise in living standards; thus sustainable development can only be pursued if population size and growth are in harmony with the changing productive potential of the ecosystem.

Yet in the end, sustainable development is not a fixed state of harmony, but rather a process of change in which the exploitation of resources, the direction of investments, the orientation of technological development, and institutional change are made consistent with future as well as present needs. We do not pretend that the process is easy or straightforward. Painful choices have to be made. Thus, in the final analysis, sustainable development must rest on political will.

(Excerpted from Foreword *From One Earth to One World: An Overview by the World Commission on Environment and Development*)

Vocabulary

sustainable development 可持续发展

World Commission on Environment and Development 世界环境与发展委员会（WCED）

United Nations General Assembly 联合国大会

Reflection and Activities

In the chapter II, "Common Challenges", *Our Common Future* lists the challenges we need to face to in 1987. Find how to describe those challenges in the book and consult the sources for information of solution in China.

Common Challenges	Descriptions in *Our Common Future*	Solutions in China
Population and Human Resources		
Food Security: Sustaining the Potential		
Species and Ecosystems: Resources for Development		
Energy: Choices for Environment and Development		
Industry: Producing More with Less The Urban Challenge		

第四节 《我们共同的未来》

导读

你知道什么是可持续发展吗？

——人类有能力使发展具有可持续性，在确保发展满足当前需要的同时又不损害子孙后代满足自身需要的能力。

《我们共同的未来》是世界环境与发展委员会关于人类未来的报告。1987年2月在日本东京举行的第八届世界环境与发展委员会通过了这项计划。1987年4月，在联合国大会上进行辩论后，《我们共同的未来》于1987年4月正式出版。

格罗·布伦特兰（Gro Brundtland）的《我们共同的未来》（1987）首次提出了可持续发展的关键概念。这本书的出版标志着我们历史上的一个重要开端。

这份报告极大地改变了国际政治思想，促使人们认识到我们应该——负责任地使用自然资源，造福子孙后代。

该报告发表后立即受到了许多积极关注，并被牛津大学出版社誉为"十年来关于世界未来的最重要文件"。事实上，可以通过宣称该报告是二十世纪最有影响力的文本来进一步阐述这一点。

在布伦特兰夫人的领导下，世界环境与发展委员会的21名成员系统地研究了人类面临的主要经济、社会和环境问题。他们以"可持续发展"为指导思想，从保护和开发环境资源、满足当代和子孙后代的需要的角度，提出了一系列政策目标和行动建议。

该报告将环境与发展结合为一个整体。报告提出的指导思想对各国政府和人民制定政策具有积极而重要的借鉴意义。

《我们共同的未来》指出，世界各国政府和人民今后必须对经济发展和环境保护两大问题承担历史责任，制定和实施正确的政策。生态环境已受到严重破坏，必须立即采取行动。

该报告首次全面概述了世界人口快速增长、粮食短缺、城市人口增长、环境恶化和气候变化等重大问题。该报告影响了当今的决策过程。报告中提到的应对气候变化等重大挑战，在今天仍然具有现实意义。

节选

可持续发展

人类有能力实现可持续发展——在确保发展满足当前需要的同时又不损害子孙后代满足自身需要的能力。可持续发展的概念确实意味着限制——不是绝对的限制，而是技术和社会组织的现状对环境资源以及生物圈吸收人类活动影响的能力所施加的限制。但技术和社会组织都可以得到进一步的管理和改进，为新的经济增长时代让路。委员会认为，普遍的贫困不再是不可避免的。贫困本身不仅是一种罪恶，可持续发展要求满足所有人的基本需求，并把机会提供给所有渴求美好生活的人。一个贫穷盛行的世界总是容易发生生态灾难和其他灾难。

满足基本需求不仅需要为大多数贫困国家创造一个新的经济增长时代，还需要确保这些贫困国家获得维持这种增长所需的公平资源份额。确保公民有效参与决策的政治制度以及国际决策中更大的民主将有助于实现这种公平。

全球可持续发展要求那些更富裕的人采用一种生活方式在地球生态承受范围内——例如在能源使用方面。此外，快速增长的人口会增加资源压力，减缓生活水平的提高；因此，只有人口规模和增长与生态系统不断变化的生产潜力相协调，才能追求可持续发展。

但归根结底，可持续发展并不是一种固定的和谐状态，而是一个变化的过程，在这个过程中，资源的开发、投资方向、技术发展方向和制度变迁都与未来和当前的需求相一致。我们并不自认为这是一个简单容易的过程。必须做出痛苦的选择。因此,归根结底,可持续发展必须取决于政治意愿。

（节选自前言《从一个地球到一个世界：世界环境与发展委员会概览》）

思考题

《我们共同的未来》第二章"共同的挑战"列出了1987年我们需要面对的挑战，了解书中对于各种环境问题的描述，并收集中国解决方案。

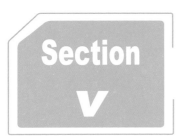

Section V Environmental Management in China

90-year-old Qu Geping, August 2020, Beijing

Introduction

Professor Qu Geping is a world-renowned environmental scientist. With an established career in environment, he has held many prominent positions, such as Deputy Director of the General Office of Environmental Protection Leading Group of the State Council, Director of the Environmental Protection Bureau of the Ministry of Urban and Rural Construction and Environmental Protection, Chairman of the Environmental and Resources Protection Committee of NPC (National People's Congress), Director of CSES (Chinese Society for Environmental Sciences), Director of CEPF (China Environmental Protection Foundation).

Many firsts in the achievements of environmental protection in China:

1. First Chinese Representative to the United Nations Environment Programme;
2. First Chief of the National Environment Protection Bureau;
3. China's first winner of the United Nations Environmental Gold Medal;

4. China's first winner of the United Nations Top Environmental Award in Tichuan;

5. The first Chinese government minister to have his work published by an international organization;

6. The first chairman of Environment and Resources Protection Committee of National People's Congress of China;

7. China's first winner of Dutch "Golden Ark" award and the highest level of the award "Conductor Award".

Professor Qu Geping is not only a pioneer in China's environmental management and environmental legal system, but also a pioneer in China's environmental education. In 1977, under the guidance of Professor Qu Geping, Tsinghua University created the first environmental engineering major in China. In 1979, he founded a secondary vocational school, Changsha Environmental Protection School. In 1981, he founded an adult training school, Qinhuangdao Environmental Protection Cadre School (renamed Environmental Management College of China in 1989, now Hebei University of Environmental Engineering), and served as the first president from 1986 to 1993. In July 2004, Mr. Qu Geping personally funded the establishment of the "Qu Geping Scholarship" in Environmental Management College of China (now Hebei University of Environmental Engineering) and Shandong University, aiming to encourage college students to study diligently, love environmental protection, and contribute to China's environmental protection cause. Hebei University of Environmental Engineering is located in Beidaihe District, Qinhuangdao City, Hebei Province. It is one of the earliest universities in China to carry out ecological environment education. At present, it is the only university in China with the characteristics of ecoenvironment education. The school has been adhering to the motto of Professor Qu Geping's inscription "Unity, Rigor, Truth-seeking and Innovation" and committed to cultivating more outstanding talents for the national ecological environmental protection cause. Guided by Xi Jinping Thought on Ecological Civilization and jointly designed by teachers and students, the Qu Geping Environmental Culture Center was inaugurated in the school in 2021.

Environmental Management in China was published by the United Nations Environment Programme in foreign languages, and Dr. Torba, Director of the United Nations Environment Programme and an expert on environmental protection, prefaced the book, "Mr. Qu Geping took the first step in the right direction of improving China's environment." The book won the "Fourth China Book Award".

Chapter I 第一章

河北环境工程学院官网

Excerpts

Part One Environmental Management Theory

I. THE GUIDING PRINCIPLES FOR ENVIRONMENTAL PROTECTION

(1) We must adhere to the policy of synchronous planning, implementation and development of economic growth, urban and rural construction and environmental improvement. The economic, social, and environmental costs and benefits shall become the criterion of quality of projects. It is recognized that to focus on either economic or environmental benefits in isolation is misguided. Proper policy takes a wholistic approach and at once consider both economic and environmental costs and benefits. This approach to integrating the three benefits represents a new contribution to environmental theory. It is indicative of the improvement in our guiding principles and has received praise both at home and abroad. It should be noted that socialist institutions have provided favorable conditions for implementing the policy of three synchronization.

(2) In establishing environmental goals and standards, it is necessary to be realistic and adopt policies which are viable within our economic limitations. This, too, should become one of our guiding principles in environmental protection work. To create a beautiful living environment is the goal of our preparations and efforts. So, to reach this goal, we must begin from the practical realities of the situation and advance one step at a time. In the past, some standards we set for ourselves were too high. The elimination of smog

would require installing huge amounts of equipment which is currently economically unfeasible. The goal to eliminate and prevent water pollution was similarly hampered. Which are set too high can't be met, or if met, the extra strain will adversely affect economic development. Thus environmental standards must be in accordance with our nation's economic reality. Once our economic standards have been improved, environmental standards can be raised accordingly. Given the vast size of our nation, naturally regional economic differences exist. In regions where economic conditions are good and pollution problems are serious, it is both permitted and realistic to impose stricter environmental standards.

(3) At present, environmental protection work is focused on management. China's environmental problems are serious. Each year losses attributed to environmental pollution and destruction total several billion U.S. dollars. How should this be solved? Two options exist. One is reliance on extensive capital investment by the central government. In this approach, proper treatment of polluted water alone would require 1.7 billion U.S. dollars. That is clearly beyond the means of the state, and thus this approach is unfeasible. The second approach is to strengthen environmental management and control environmental pollution through better management. Currently this is a more economical and effective approach to solving environmental problems. We presented these two general approaches and means in a report to the State Council. The leaders of the State Council favored the second approach. The purpose of establishing the Environmental Protection Commission and approval for establishment of the National Environmental Protection Agency is to strengthen environmental management. Experience has shown that many problems of environmental pollution can be solved by strengthening and improving environmental management. And just what does the broad term "management" involve? The State Council has stipulated four tasks required to augment the work of environmental protection. They are: planning, coordination, supervision and guidance.

(Excerpted from *Environmental Management in China,* Qu Geping, United Nations Environment Programme and China Environmental Science Press, First Edition 1991)

Reflection and Activities

At an international conference held by the United Nations Environment Programme in August 1990, Qu Geping briefed Dr. Torba on the eight systems and measures of environmental management that China was widely promoting, and Torba was extremely excited. He told the head of the United Nations Environment Programme: "This is a very correct system and measure, many of which he has long wanted to do, such as the environmental impact assessment system, the environmental target responsibility system, the city quantitative assessment, and the sewage charging system, but most countries have not done it yet. It is China that has developed environmental management. China's foreign policies and a series of systems and measures are creative and are ahead of the world." Torba highly praised Qu Geping as an "unremitting warrior" on the environmental protection front.

1. Please find out the measures for environmental management in China according to the excerpt.

2. Suppose you were invited to introduce Qu Geping, the pioneer of environmental protection in China, to a program called Speak to the world in CCTV-4 as a story teller. This program focuses on telling China's story and spreading China's voice. What will you tell and how do you understand "Qu Geping is an unremitting warrior"?

第五节 5

《中国环境管理》

导读

曲格平教授是世界知名的环境科学家。在环境领域曾担任过许多重要职务,如国务院环境保护领导小组办公厅副主任,城乡建设和环境保护部环境保护局局长,全国人大环境与资源保护委员会主席。中国环境科学学会理事,中国环境保护基金会理事。

中国环保成就的多项第一:

1. 首位中国驻联合国环境规划署代表;
2. 首任国家环境保护局局长;
3. 中国首位荣获联合国环境金质奖章;
4. 中国首位在四川获得联合国最高环境奖;
5. 第一位在国际组织发表著作的中国政府部长;
6. 全国人民代表大会环境与资源保护委员会首任主席;
7. 中国首位荣获荷兰"金方舟"奖和最高级别的"指挥家奖"。

曲格平教授不仅是中国环境管理和环境法制的开拓者,也是中国环境教育的开拓者。1977年,在曲格平教授的指导下清华大学创建了全国第一个环境工程专业。1979年创办一所中专学校——长沙环境保护学校。1981年创办一所成人高等院校——秦皇岛环境保护干部学校(1989年改称中国环境管理干部学院,现河北环境工程学院),1986—1993年兼任第一任院长。为了激励有志青年投身国家环境保护事业,2004年7月曲格平先生个人出资在中国环境管理干部学院(现河北环境工程学院)和山东大学设立了"曲格平奖学金",旨在鼓励大学生勤奋学习,热爱环保,为我国的环保事业做贡献。河北环境工程学院位于河北省秦皇岛市北戴河区,是我国最早开展生态环境教育的高校之一。目前,它是中国唯一一所以生态环境教育为办学特色的本科大学。学校一直秉承曲格平教授题词"团结、严谨、求实、创新"的校训,致力于为国家生态环境保护事业培养更多的优秀人才。以习近平生态文明思想为指导,在师生共同设计下,曲格平环境文化馆于2020年在学校落成。

《中国环境管理》由联合国环境规划署组织出外文版,联合国环境规划署主任、环境保护专家托尔巴博士为该书作序,"曲格平先生在向改善中国环境方面的正确方向迈出了第一步"。该著作获"第四届中国图书奖"。

节选

环境保护理论

一、环境保护指导原则

1. 要坚持经济增长、城乡建设和环境改善同步规划、实施、发展的政策。经济、社会、环境成本和效益应当成为项目质量的标准。人们认识到，孤立地关注经济或环境惠益是错误的。我们应当用适当的政策，并采取全面的方法，同时考虑经济和环境成本和效益。这种整合了这三种优势的方法代表了对环境理论的新贡献，表明了我们指导原则的改进，并得到了国内外的好评。应当看到，社会主义制度为贯彻三步并进提供了有利条件。

2. 在制定环境目标和标准时，必须采取在经济限制范围内现实可行的政策。这也应成为我们环境保护工作的指导原则之一。创造美好的生活环境是我们准备达成并为之努力的目标。因此，要实现这一目标，我们必须从实际情况出发，一步一个脚印地前进。过去，我们为自己设定的一些标准太高了。消除雾霾需要安装大量设备，这在经济上是不可行的。消除和防止水污染的目标也同样受到阻碍。那些定得太高是无法满足的，或者如果满足，额外的压力将对经济发展产生不利影响。因此，环境标准必须符合我们国家的经济现实。一旦我们提高了经济标准，环境标准就可以相应地提高。鉴于我们国家的庞大规模，自然存在区域经济差异。在经济条件良好而污染问题严重的地区，实施更严格的环境标准既是被允许的，也是现实可行的。

3. 目前，环境保护工作以管理为重点。中国的环境问题很严重。每年由于环境污染和破坏造成的损失总计数十亿美元。应该如何解决这个问题，存在两种选择。一个是依赖中央政府的广泛资本投资。在这种方法中，仅对受污染的水进行适当的处理就需要 17 亿美元。这显然超出了国家的能力范围，因此这种方法是不可行的。第二种方法是通过更好的管理来加强环境管理和控制环境污染。目前，这是解决环境问题的一种更经济有效的方法。我们在提交国务院的报告中提出了这两种一般方法和手段。国务院领导人赞成第二种办法，成立环境保护委员会和批准设立国家环境保护厅，目的是加强环境管理。经验表明，通过加强和改善环境管理，可以解决许多环境污染问题。广义的术语"管理"究竟涉及什么？国务院规定了加强环境保护工作所需的四项任务。它们是：规划、协调、监督和指导。

（节选自《中国环境管理》，曲格平，联合国环境规划署、中国环境科学出版社，1991年第一版）

思考题

在 1990 年 8 月一次联合国环境规划署召开的国际会议上，曲格平向托尔巴博士介绍了中国正在广泛推行的环境管理的八项制度和措施。托尔巴异常兴奋，他向在座的联合国环境署的主管官员说："这是非常正确的制度和措施，许多是我长期想做的，如环境影响评价制度、环境目标责任制、城市定量考核、排污收费制度好，但是，现在大多数国家还没有做。是中国推进了环境管理发展。中国环境管理的外政策和一系列的制度、措施是富有创造性的，走在了世界的前边。"托尔巴高称赞曲格平是环境保护战线上一名"不懈的战士"。

1. 请根据上面《中国环境管理》的部分节选找出中国环境管理的措施。

2. 假设您受邀将在 CCTV-4 的一个名为"告诉世界"的节目上介绍中国环保的开拓者——曲格平。这个节目的重点是讲述中国故事，传播中国声音。你将如何介绍？你如何理解"曲格平是一个不懈的战士"？

Section VI — China's Road of Green Development

解振华视频发表获奖感言

解振华与河北环境工程学院同志合影

Introduction

Xie Zhenhua, born in 1949, was the Special Representative for Climate Change Affairs of China since April 2015. He successively served as former Vice Chairman of National Development and Reform Commission, former Minister of State Environmental Protection Agency; Vice Chairman of the 12th Population, Resources and Environment Committee of the CPPCC, Member of the 16th and the 18th central Committee of CPC, member of the 15th and the 17th Central Commission for Discipline Inspection of CPC, and president of the Environmental Management college of China (Hebei University of Environmental Engineering), promoting the development of environmental education in China.

Awards: GEF Global Leadership Award (2002); UNEP Sasakawa Environment Prize and World Bank Special Green Award (2003); Alliance to Save Energy's Energy Efficiency Visionary Award (2009); WWF Livable Planet Leadership Award (2015); Lui Zhiwo Award for Sustainable Development (2017).

Xie Zhenhua has taken charge of environmental protection, resource conservation, energy saving, pollution reductions and climate change affairs in China. As head of Chinese negotiation delegation to the UN Climate Conferences since 2007, he played an important role in adoption, signing and entry into force of the Paris Agreement as well as the adoption of Paris Agreement implementation rules. He has promoted cooperation on energy conservation, emission reductions and low-carbon development between China and 30 countries, regions and international organizations, and helped train 1,000 officials and technicians working in the field of climate change for 100 developing countries.

Excerpts

The Action of China Contributing to Global Green Development

While solving domestic issues on green development, China has always made its own contribution to global sustainable development. From the international perspective, since the founding of the PRC, China has been actively involved in international development cooperation. It has provided assistance worth 400 billion yuan to 166 countries and international organizations, and dispatched more than 600,000 aid workers, making important contribution to developing countries in realizing the Millennium Development Goals. In May 2017, the Chinese Ministry of Environmental Protection released the "Belt and Road Ecological and Environmental Cooperation Plan". The Plan points out that in the field of Eco-environmental cooperation, China has actively deepened multilateral and bilateral dialogue, exchanges and cooperation with countries along the Silk Road Economic Belt and 21st century Maritime Silk Road, strengthened information support services for the ecological environment, and promoted cooperation in environmental standards, technology and industry, and has made positive progress and obtained satisfactory results.

In terms of bilateral green governance, China has clearly and definitely made its first commitment to international assistance in the national plan. During the process of developing the 13th Five Year Plan, China proposed that it will improve the way and enlarge the scale of foreign assistance; provide developing countries with more free consultations and training in human resources, development planning and economic policy; expand foreign cooperation and assistance in fields such as science and technology, education, medical and health care, disaster prevention and reduction, environmental improvement, wildlife protection and poverty alleviation; and increase humanitarian assistance.

At the multilateral level, China provides impetus for the international multilateral agendas, focused on the field of global climate management. China fully exerts its influence as a big country and plays a key role in promoting the Paris Agreement on schedule. Recently, China has issued joint statements successively with the United Kingdom, the United States, India, Brazil, the European Union and France to form a series of consensuses on cooperation against climate change and promotion of multilateral agendas. The consensuses in the China-US as well as China-France joint statements in particular became the basis for compromise during the final stage of the negotiations for the Paris Agreement. During the Paris Conference, the Chinese delegation, with a Responsible and constructive attitude, fully participated in negotiations on various issues, intensively carried out shuttle diplomacy, and supported and collaborated with host country France and the United Nations on relevant work.

China's vigorous promotion of South-South cooperation is also an important aspect of China's participation in multilateral green governance. In view of the fact that many developing countries arc lagging behind in their economy and infrastructure, and are vulnerable to the disadvantages of climate change, China has for many years provided active and effective support for African countries, small- island counties and the least-developed countries in improving their capabilities to cope with climate change by carrying out South-South cooperation on climate change. Since 2011, China has invested 410 million yuan to help countries to improve their infrastructure and strengthen their capacity for coping with climate change.

China is taking actions to participate in the development of the 2030 Agenda. Since the start of the development of the 2030 Agenda, China has actively taken part and collaborated in the discussions and consultations of the United Nations, widely listened to the views of various parties in China, and made many useful comments on the agenda. With the support of the United Nations Development Program, the China United Nations Association held three informal national-level consultations respectively in Beijing, Kunming and Beijing in November and December 2012 and March 2013 so as to listen extensively to opinions of all walks of life about the Post-2015 Development Agenda. Among those who Participated, 75 percent came from social groups. The consultations focused on the six major areas of poverty reduction and inclusive growth, environmental protection and green development policy, global health, women and children, education and international cooperation.

On the regional level, China will participate in the process of global green governance via the Belt and Road Initiative. This initiative will be fully dependent on the existing bilateral and multilateral mechanisms between China and related countries and draw support from existing regional cooperation platforms. The experiences of China in environmental governance over the years will bring benefits to many countries.

(Excerpted from *China's Road of Green Development*, Xie Zhenhua & Pan Jiahua)

Vocabulary

poverty alleviation 缓解和消除贫困
climate-change mitigation 减缓气候变化
multilateral green governance 多边绿色治理
the Belt and Road Initiative "一带一路"倡议

Reflection and Activities

1. The Paris Agreement states that all parties will strengthen the global response to the threat of climate change by holding the increase in global average temperature below 2 degrees Celsius above pre-industrial levels and working to limit the rise to 1.5 degrees Celsius. The world will peak greenhouse gas emissions as soon as possible and achieve net zero greenhouse gas emissions in the second half of this century. What's the attitude of our country and what's your opinion on it?

2. As participant, contributor and leader in global ecological progress, China has made great contributions to the world. Please list contributions China has made to global environmental governance.

第六节 《中国绿色发展之路》

导读

解振华，1949年生，2015年4月起任中国气候变化事务特别代表。他曾任国家环境保护总局局长和国家发展和改革委员会副主任，在担任国家环境保护总局局长期间，兼任中国环境管理干部学院（现河北环境工程学院）院长，对中国环保教育事业的发展起了巨大的推动作用。

获奖：全球环境基金全球领袖奖（2002）；联合国环境规划署笹川环境奖和世界银行特别绿色奖（2003）；节能联盟能源效率远见奖（2009）；世界自然基金会宜居地球领袖奖（2015）；吕志和可持续发展奖（2017）。

解振华负责中国环境保护、资源节约、节能减排、气候变化等方面的工作。自2007年起担任联合国气候大会中国谈判代表团团长，为《巴黎协定》的通过、签署和生效以及《巴黎协定》实施规则的通过发挥了重要作用。推动中国与30个国家、地区和国际组织开展节能减排和绿色低碳发展合作，为100个发展中国家培训了1000名气候变化领域官员和技术人员。

节选

中国助力全球绿色发展的行动

在解决国内绿色发展问题的同时，中国始终为全球可持续发展做出自己的贡献。从国际上看，新中国成立以来，中国积极参与国际发展合作。向166个国家和国际组织提供了价值4000亿元人民币的援助，派出超过60万援助人员，为发展中国家实现千年发展目标作出了重要贡献。2017年5月，中国环保部发布了《"一带一路"生态环境保护合作规划》。《规划》指出，在生态环境合作领域，中国积极深化与丝绸之路经济带和21世纪海上丝绸之路沿线国家的多双边对话、交流与合作，加强生态环境信息支撑服务，推进环境标准、技术和产业合作，取得积极进展，取得了令人满意的成果。

在双边绿色治理方面，中国首次在国家规划中明确、明确地作出了国际援助承诺。在制定"十三五"规划过程中，中国提出改进对外援助方式，扩大对外援助规模；在人力资源、发展规划和经济政策方面为发展中国家提供更多免费咨询和培训；扩大在科技、教育、医疗卫生、防灾减灾、环境治理、野生动物保护、扶贫等领域的对外合作与援助。并增加人道主义援助。

在多边层面，中国推动国际多边议程，重点关注全球气候管理领域。中国充分发挥大国影响，为如期推进《巴黎协定》发挥关键作用。近期，中方先后与英国、美国、印度、巴西、欧盟、法国等发表联合声明，就合作应对气候变化、推进多边议程形成一系列共识。尤其是中美、中法联合声明中的共识，成为巴黎协定谈判最后阶段妥协的基础。巴黎会议期间，中国代表团本着负责任和建设性的态度，全面参与各项议题谈判，深入开展穿梭外交，并与东道国法国和联合国就有关工作给予支持和配合。

中国大力推进南南合作，也是中国参与多边绿色治理的重要方面。鉴于许多发展中国家经济落后、基础设施落后、易受气候变化不利影响，中国多年来为非洲国家、小岛屿国家和最不发达国家提供了积极有效的支持。发达国家开展气候变化南南合作，提高应对气候变化能力。2011年以来，中国已投入4.1亿元人民币，帮助各国改善基础设施，增强应对气候变化能力。

中国正在采取行动参与2030年议程的制定。2030年议程制定以来，中国积极参与和配合联合国的讨论和磋商，广泛听取中国各方意见，对议程提出许多有益意见。在联合国开发计划署的支持下，中国联合国协会分别于2012年11月、12月和2013年3月在北京、昆明和北京举行了3次国家级非正式磋商，广泛听取各界关于2015年后发展议程的意见。在参与的人中，75%来自社会群体。磋商围绕减贫与包容性增长、环境保护与绿色发展政策、全球健康、妇女儿童、教育与国际合作六大领域展开磋商。

在区域层面，中国将通过"一带一路"倡议参与全球绿色治理进程。这一倡议将充分依托中国与有关国家现有的双边和多边机制，并得到现有区域合作平台的支持。中国多年来在环境治理方面的经验将惠及许多国家。

（节选自《中国的绿色发展之路》解振华，潘家华，外文出版社）

思考题

1.《巴黎协定》规定，各方将加强全球应对气候变化威胁的力度，将全球平均气温上升幅度控制在工业化前水平以上2摄氏度以内，并努力将上升幅度控制在1.5摄氏度以内。世界将尽快达到温室气体排放峰值，并在本世纪下半叶实现温室气体净零排放。我国对此持什么态度，你对此有何看法？

2.作为全球生态文明进步的参与者、贡献者和引领者，中国对全球环境治理作出了巨大贡献。请列举中国对全球环境治理的积极贡献。

Policies and Actions on Sustainable Development in China

Introduction

Zhang Kunmin, born in 1936, is a doctoral supervisor. He is currently the general advisor of the State Environmental Protection Administration, the secretary general of International Cooperation Committee on Environment and Development in China, and the vice chairman of the Chinese Society for Sustainable Development. Zhang Kunmin is a doctoral supervisor at Tsinghua University and Renmin University of China, and an adjunct professor at 10 other universities including Peking University, Nanjing University, Fudan University, and the National School of Governance. Since 1985, he has been transferred to the executive vice president of Environmental Management College of China (Hebei University of Environmental Engineering now) for 3 years.

The main research areas of Professor Zhang Kunmin are sustainable development theory and practice, low-carbon economy, environmental policy and law, and international cooperation. He has published 19 books, among which, he edited "Introduction to Low Carbon Economy"(2008), "Eco-city Evaluation and Indicator System"(2003), "Twenty Years of Environmental Protection Administration in China"(1994), "Report on Environmental Protection Investment in China"(1992), "Speech on Environmental Protection Law"(Chinese version 1990, English version 1992, Japanese version 1994), etc. His monographs include

"Policies and Actions on Sustainable Development in China"(English version 2001, Chinese version 2004). 20 translation and proofreading books include "Asian Environment White Paper Volume 1"(Japan, 2005), "Encyclopedia of the American Environment"(U.S., 2003), "Expanding the Means of Measuring Wealth: Indicators of Environmental Sustainability"(World Bank, 1998), etc. He has published more than 100 papers and reports, and has made in-depth research and outstanding achievements in his area.

Excerpts

Sustainable development theory holds that development and environment together form an organic whole. The Rio Declaration stressed that environmental protection must be a part of the developmental process. This theory consists of the following elements:

1. Sustainable development does not reject economic growth, especially in poor nations. It is necessary, however, to review the approaches to economic growth. To achieve growth in a sustainable sense, it is necessary to transform the mode of production from an extensive to an intensive model, to reduce the environmental pressure caused by a unit of economic activities, and to study and resolve economic distortions and fallacies. Since environmental degradation results from the economic process, it is logical to find the solution to environmental degradation within the economic process.

2. Sustainable development is based on natural assets and is compatible with nature's ability to sustain economic growth. Sustainability can be achieved through appropriate economic means, technical measures and government intervention. Efforts should be focused on reducing the depletion rate of natural assets to a level lower than the recycling rate of resources. In addition, clean techniques and sustainable modes of consumption should be encouraged so that the amount of wastes caused by a unit of economic activities can be reduced to the minimum level.

3. Sustainable development aims at improving the quality of life, a goal in harmony with social progress. Economic development is a concept far more intensive than economic growth. Economic growth is generally defined as a rise of per capita gross national product (GNP). Development, on the other hand, necessitates progress in social and economic structures, in particular the realization of a series of goals set for social development.

4. Sustainable development furthermore, recognizes and entails the value of environmental resources. Such value is manifested not only in the fact that the environment sustains and serves the economic system, but also in the fact that it nurtures the entire life support system. Therefore, it is necessary to include input of environmental resources in production and related services into the cost of production and pricing of products. Eventually, the accounting system of a national economy has to be modified.

5. Implementation of the sustainable development strategy is conditional on the establishment of applicable policies and legal systems, particularly "comprehensive decision-making" and "public involvement". The old practice of one department single-handedly formulating and implementing economic, social and environmental policies is no longer viable. Instead, decisions should be based on thorough social, economic and environmental considerations, scientific principles, maximum information and comprehensive requirements. The principle of sustainable development should be incorporated into major legislation and decision-making processes involving economic development, population, environment, resources and social security.

(Excerpted from *Policies and Actions on Sustainable Development in China*, Zhang Kunmin, China Science Press)

Vocabulary

sustainable development 可持续发展

sustainablity 可持续性

per capita gross national product (GNP) 人均国民生产总值

comprehensive decision-making 综合决策

public involvement 公众参与

Reflection and Activities

1. List at least one classic case of environment and sustainable development in the 1980s and 1990s in

China.

2. Interview 5–6 students on campus to find out how they transform the concept of green and low-carbon into acts in daily life.

3. Report to the class your findings and analyses. Give suggestions as to the practical actions to the construction of an ecological space with blue sky, green earth and clear water.

《中国可持续发展政策与行动》

导读

张坤民，1936 年生，博士生导师，现任国家环保局总顾问、中国环境与发展国际合作委员会秘书长，兼任中国可持续发展研究会副理事长，任清华大学和中国人民大学的博士生导师，北京大学、南京大学、复旦大学、国家行政学院等其他 10 所大学的兼职教授，1985 年起调任中国环境管理干部学院（现河北环境工程学院）常务副院长 3 年。

张坤民教授的主要研究领域是可持续发展理论与实践、低碳经济、环境政策和环境法、国际合作等。他出版著作 19 部，主编《低碳经济论》（2008）、《生态城市评估与指标体系》（2003）、《中国环境保护行政二十年》（1994）、《中国环境保护投资报告》（1992）、《环境保护法讲话》（中文版 1990、英文版 1992、日文版 1994）等；专著《关于中国可持续发展的政策与行动》（英文版 2001、中文版 2004）。译校书籍 20 种，包括：主译审《亚洲环境白皮书第一卷》（日本、2005）、《美国环境百科全书》（美国、2003）、《扩展衡量财富的手段：环境可持续发展的指标》(世界银行、1998) 等。发表论文与报告 100 余篇。

节选

可持续发展理论认为，发展与环境是一个有机的整体。《里约宣言》强调环境保护必须成为发展进程的一部分。该理论由以下要素组成：

1. 可持续发展并不排斥经济增长，尤其是在贫穷国家。然而，有必要审视经济增长的方式，实现可持续增长，将生产方式从粗放型转变为集约型，降低单位经济活动对环境造成的压力，研究和解决经济扭曲和谬误。由于环境退化是经济过程的结果，因此在经济过程中找到环境退化的解决方案是合乎逻辑的。

2. 可持续发展以自然资产为基础，与自然维持经济增长的能力相适应。可持续发展可以通过适当的经济手段、技术措施和政府干预来实现。应重点努力将自然资产的消耗率降低到低于资源回收率的水平。此外，应鼓励清洁技术和可持续消费方式，将单位经济活动产生的废物量降至最低水平。

3. 可持续发展旨在提高生活质量，与社会进步相协调。经济发展是一个比经济增长更深入的概念。经济增长通常被定义为人均国民生产总值（GNP）的增长。另一方面，发展需要社会和经济结构的进步，特别是实现为社会发展设定的一系列目标。

4.可持续发展进一步承认环境资源的价值。这种价值不仅体现在环境对经济系统的维持和服务，而且还体现在它滋养了整个生命支持系统。因此，有必要将生产和相关服务中的环境资源投入纳入产品的生产成本和定价。最终，必须修改国民经济的统计体系。

5.可持续发展战略的实施取决于建立适用的政策和法律制度，特别是"综合决策"和"公众参与"。一个部门单独制定和实施经济、社会和环境政策的旧做法已不可行。相反，决策应该基于全面的社会、经济和环境考虑、科学原则、最大信息和综合要求。将可持续发展原则纳入经济发展、人口、环境、资源和社会保障等重大立法和决策过程。

（节选自《中国可持续发展政策与行动》，张坤民，中国科学出版社）

思考题

1.列举中国在二十世纪八、九十年代环境与可持续发展的经典案例。

2.在校园里采访5—6名学生，了解他们如何将绿色和低碳的概念转化为日常生活中的行为。

3.向全班报告你的发现和分析并就建设天蓝、地绿、水清的生态空间的实际行动提出建议。

Section viii: General Theory of Environmental Protection

Introduction

Professor Liu Tianqi (1927−2011) was one of the pioneers of Environmental Science in China. He once participated in the establishment of the discipline of Environmental Management, which is characterized by urban environmental planning and management. Serving as director and deputy secretary general of Chinese Society for Environmental Sciences and vice president of environmental education branch, he made important contributions to the establishment and development of Chinese Society for Environmental Sciences. In 1985, he was transferred to Environmental Management College of China as the vice president. Professor Liu Tianqi once served as deputy director of the editorial board of Encyclopedia of China·Environmental Science Volume and chief editor of Environmental Engineering. He also edited *General Theory of Environmental Protection*, *Introduction to Environmental Protection* and other works.

General Theory of Environmental Protection is a comprehensive basic textbook of environmental protection compiled by the Environmental Management College of China, organized by domestic well-known experts, scholars and experienced environmental protection cadres under the direct leadership of the National Environmental Protection Administration in the mid-1980s. Since its publication, the book has played a very important role in the training and business assessment of China's environmental protection cadres.

Professor Liu Tianqi took the lead in developing the research on the norms and methods of urban environmental planning. He is the earliest expert in developing the norms and methods of urban environmental planning in China. His pioneering work in this field has brought a new situation to the study of urban environmental planning norms in China. In the process of exploring urban environmental planning, Professor Liu Tianqi has formed his own academic thought which has the following characteristics:

1. It is necessary to take economic construction as the center, promote the close integration and overall coordination of urban environmental planning with economic and social development planning and overall urban construction planning, and strive to integrate urban environmental planning into urban economic and social development planning.

2. The scientific and practical method of urban planning personnel is based on the concise and practical stage of urban planning with Chinese characteristics, and the scientific and practical method of urban planning personnel is to be in line with the reality of China.

3. The comprehensive environmental planning method is adopted, and the opinions of various departments and industries are absorbed and adopted to ensure the feasibility of the planning scheme.

4. Due to China's vast territory and complex situation, cities in different regions and types should highlight their own characteristics, formulate targeted urban environmental planning, and strive for a virtuous cycle of urban ecosystem.

Excerpts

The environmental protection strategy of urban macro environmental planning mainly includes: strategic focus, strategic objectives and strategic countermeasures. It is the basis for formulating macro environmental planning scheme.

Based on the analysis of urban ecological characteristics, the comprehensive carrying capacity of resources and environment, the coordination degree of economy and environment, and the main environmental problems of the city, the strategic focus of environmental protection is determined. According to the current situation of China's urban environment, there are generally the following three strategic priorities. First, we should focus on rationally developing, utilizing and protecting water resources. The shortage of water resources is a worldwide problem. 1.2 billion people all over the world can't drink clean water; 300 cities in China are short of water, of which more than 50 cities are seriously short of water. Therefore, the issue of water resources should be the preferred strategic focus. Second, it is the control and management of air pollution. Coal smoke air pollution is a common environmental pollution problem in China's cities. The mortality of lung cancer and respiratory diseases shows an obvious upward trend, and the emergence of "Oxygen Bar" in cities shows the seriousness of this problem. Therefore, improving energy efficiency, improving energy structure and taking effective measures to control air pollution should be the strategic focus. Third, in the process of sustained and rapid economic development, we should figure out how to make a high degree of material civilization coexist with a clean, comfortable and beautiful environment.

Aiming at the strategic goal of "controlling the environmental pollution in Qinhuangdao city in 2010", the following countermeasures are put forward.

First, total quantity control should be carried out according to environmental function zoning, and key pollution sources should be controlled in batches. Establish various control areas or pollution control units, such as smoke control area, noise control area and so on. Implement the sewage permit system.

Second, formulate and implement the implementation of cleaner production programs. First of all, it is necessary to improve the understanding of economic workers and environmental protection workers on cleaner production, and understand that cleaner production is two whole process control (resource utilization, the whole process of processing and the whole life cycle of products).

Third, increase investment in environmental protection and improve the comprehensive benefits of environmental protection measures. Incorporate investment in environmental protection (the cost of preventing and controlling pollution) into the government budget. We will actively expand funding channels, establish environmental protection funds, and raise the investment ratio for environmental protection.

(Excerpted from *General Theory of Environmental Protection*, Chapter XVI, Section III, Liu Tianqi, China Environmental Press, Second Edition, 1997.9)

Vocabulary

macro 宏观的

comprehensive 综合的

countermeasures 对策

incorporate 合并的

Reflection and Activities

1. Compared with rural environmental governance, what are the unique characteristics of urban environmental planning?

2. Considering the prevailing concern about municipal wastes, what can we do to solve this problem?

3. On April 8, 2021, China Environmental Protection Foundation (CEPF) and its partners jointly launched the public welfare activity "U+School Meal Plan", which aims to help rural school students improve the nutritional health, food safety and reduce waste of school meals. At the same time, it guides students to eat a balanced diet, develop the good habits of being diligent, thrifty and cherishing food, and build a resource-saving campus. Suppose you are planning to propose to CEPF another activity which could improve the college campus environments. Name the activity and write a letter to CEPF to introduce your programme.

第八节 《环境保护通论》

导读

刘天齐教授是我国环境科学开创者之一,曾参与创建了环境管理学科,其学科特色为城市环境规划管理。刘天齐教授曾任中国环境科学学会理事、副秘书长,环境教育分会副会长等职,为我国环境科学学会的创立与发展作出了重要贡献。1985 年调入中国环境管理干部学院任副院长。刘天齐教授曾任《中国大百科全书·环境科学卷》编委会副主任及环境工程部分主编,主编了《环境保护通论》《环境保护概论》等著作。

《环境保护通论》是 20 世纪 80 年代中期在国家环境保护局直接领导下,由中国环境管理干部学院组织国内知名专家、学者和富有实践经验的环保干部共同编写的一本综合性环境保护基础教材。该书出版以来对我国环保干部的培训与业务考核等起到了十分重要的作用。

刘天齐教授率先开拓了城市环境规划规范及方法的研究,是我国开拓城市环境规划规范及方法研究最早的专家。他在这个领域的开创性工作,使我国城市环境规划规范的研究出现了新的局面。刘天齐教授在探索城市环境规划的过程中,形成了自己的学术思想,并有以下几个方面的特点:

1. 以经济建设为中心,促使城市环境规划与经济社会发展规划、城市建设总体规划紧密结合、统筹协调,并力求将城市环境规划纳入城市经济社会发展规划。

2. 从中国城市实际出发,力求符合中国特色社会主义初级阶段的现实,方法科学、合理、简明、实用,参与人员是科技人员与管理人员相结合,因此制定的城市环境规划可操作性强,具有显著的实用性。

3. 采用综合环境规划法,注意吸收和采纳各部门、各行业的意见,保证了规划方案的可行性和可操作性。

4. 由于中国国土辽阔,情况复杂,因此不同区域、不同类型的城市都要突出自身的特点,有针对性地制定城市环境规划,力求城市生态系统良性循环。

节选

城市宏观环境规划的环境保护战略,主要包括:战略重点、战略目标和战略对策,它是制定宏观环境规划方案的基础。

在城市生态特征分析，资源与环境综合承载力分析，经济与环境协调度分析，以及城市主要环境问题分析等的基础上，确定环境保护战略重点。根据我国城市环境现状，大体上有下列三个战略重点。一是合理开发利用和保护水资源。水资源紧缺是世界性的问题，全世界有12亿人喝不到清洁的水；我国有300个城市缺水，其中有50多个城市严重缺水。所以，水资源问题应是首选的战略重点。二是大气污染的控制与管理。煤烟型大气污染是我国城市中较为普遍的环境污染问题，肺癌及呼吸系统疾病的死亡率呈明显上升趋势，以及城市中"氧吧"的出现就说明了这一问题的严重性。所以，提高能源利用效率，改善能源结构，采取有力措施控制大气污染应作为战略重点。三是在经济持续快速发展的过程中，如何使高度的物质文明与清洁、舒适、优美的环境并存。

针对"2010年秦皇岛市的环境污染要得到控制"的战略目标，提出了下列对策。

其一，按环境功能区划进行总量控制，分批控制重点污染源。建立各类控制区或污染控制单元。如：烟尘控制区、噪声控制小区等。实行排污许可证制度。

其二，制定并实施推行清洁生产的方案。首先要提高经济工作者和环保工作者对清洁生产的认识，懂得清洁生产是两个全过程控制（资源利用、加工的全过程及产品整个生命周期）。

其三，增加环保投入，提高环境保护措施的综合效益。把环境保护投资（防治污染的费用）纳入政府预算。积极拓宽资金渠道，建立环境保护基金，提高环境保护投资比。

（选自《环境保护通论》第十六章第三节）

思考题

1. 与农村环境治理相比，城市环境规划有哪些独特之处？

2. 考虑到人们对城市垃圾的普遍关注，我们能做些什么来解决这个问题？

3. 2021年4月8日，中华环境保护基金会与合作伙伴共同发起公益活动"U+校餐计划"，旨在帮助乡村在校学生提升校餐的营养健康、食品安全、减少浪费，同时引导学生均衡饮食，养成勤俭节约、爱惜粮食的好习惯，建设资源节约型校园。假设你正计划向中华环境保护基金会提出一项可以改善大学校园环境的活动，请为活动命名，并写一封信给中华环境保护基金会介绍你的计划。

Section ix — On the Relation between Sustainable Development of Chinese Ancient Civilization and Ecologic Environment

The picture is taken from the Hebei University of Environmental Engineering

Introduction

Kong Fande, born in April, 1945, director of the Institute of Ecology, Environmental Management College of China (now Hebei University of Environmental Engineering), is a respectable professor of Hebei University of Environmental Engineering and member of the CPPCC (Chinese People's Political Consultative Conference) Qinhuangdao Committee, has devoted his whole life to the teaching and research on ecological conservation and the research of environmental protection and development of Qinghuangdao city.

Starting his career as an environmentalist in 1982, Professor Kong Fande was engaged in the teaching and research of ecological protection. He published more than 30 papers, 10 textbooks and monographs, and presided over 20 scientific research projects. Among them, the "Guide to Urban Environmental Planning and Methods" won the third prize of the National Education Commission, the State Environmental Protection Administration, and the National Environmental Education Excellent Textbook in 1995. In 2001, "Urban Ecological Environment Construction and Protection Planning" won the "Five One Project Award" of Qinhuangdao Municipal Party Committee and Municipal Government. "Research on the Methods of Environmental Planning in Small and Medium-sized Cities" won the third prize of the State Environmental Protection Administration for Scientific and Technological Progress.

Professor Kong Fande witnessed the actual launch and development of China's environmental protection. Until his retirement in 2008, he has made great contribution to the cause by studying the relationship between environmental protection and the booming economy of Qinghuangdao. During 26 years of his career, over 30 proposals were submitted to the CPPCC Qinhuangdao Committee, and he delivered more than 10 speeches on the local CPPCC conference, which has exerted great influence on designing of the blueprint of the city, including establishing a new area in the west of Haigang district. By now it has become a reality.

Professor Kong Fande's contribution is an invaluable asset to the flourishing of ecological civilization in China.

His research and thought is mainly centered on the concept of "harmony between man and nature". The following is introduction of his understanding.

1. When it comes to ancient thought on development, it is of great significance to discuss the idea of "harmony between man and nature", which is a topic of great concern and constant renewal in Chinese philosophy. The simple sustainable development idea contained in the thought of "harmony between man and nature" still shines with the light of wisdom up to now. China is vast in land and resources and has been a great agricultural country since ancient times. In ancient times, when productivity was low, man was extremely weak to nature. The ancient sages of our country started from the awe of life and nature in the primitive religious belief, and the dependence of people on nature in the period of agricultural civilization, putting forward the idea of "harmony between man and nature", which emphasized that human behavior should be in harmony with natural law and moral reason should be consistent with natural reason.

In the modern age of advanced technology, people use science and technology to exploit natural resources in an aggressive way. This has resulted in severe damage to the balance of nature. Mankind also faces a grave existential crisis. Xi Jinping, China's president, who has been deeply nourished by traditional Chinese culture and focusing on exploring the road of Chinese ecological civilization construction in China from the historical perspective, has inherited the ecological wisdom of the ancient Chinese "harmony between man and nature" thought, and formed Xi Jinping's ecological civilization thought which pointed out the direction of ecological environment protection in the new era.

2. Traditional Chinese Confucianism has taken the thought that man is the integral part of nature as one of its main ideas, and emphasized the unity between human being and ecology. "Man is an integral part

of nature" is created in the *Zhou Changes* (《周易》), developed by Confucius (孔子), Mencius (孟子), Xun Zi (荀子), eventually established by Dong Zhongshu (董仲舒) and enlarged by Cheng Yi and Cheng Hao (二程), Zhu Xi (朱熹) and Wang Yangming (王阳明). From then on, the thought is completed and stated in a plain and direct way.

"Nature and man is unified as one" includes harmonious thought from San Cai ("三才") theory (earth and man of harmony with God). From the perspectives of modern idea of ecology, these viewpoints inspire the sense of environmental protection, reflect the thought of ecological harmony, show the codes of appropriate conservation, and give birth to the measures of environmental conservation. However, due to the limitation of historical situation, theory of "harmony between man and nature" also has its own shortcomings in terms of taking practical measures and concrete actions which is not the equivalent of modern concept.

3. The modern conversion of the thought concerns two aspects:

First, modern society inherits the practical elements from the thought of "harmony between man and nature". Our society needs to achieve the thinking of unification, to harmonize the relation of man and nature, to absorb the benevolent emotion and to care for every creature. We should use the principles of ecological balance to protect the environment and abide by the ethics of sincerity to upgrade the function of natural sources. Second, shortcomings of the thought should be dealt with so that it can be re-defined and improved in the new era.

Excerpts

Sustainable development of the Chinese ancient civilization was seldom seen in the world, leaving abundant historic experience and wisdom to the later generations. There is a variety of reasons for the sustainable development of the Chinese ancient civilization. Among them, ecological environment changes are of great importance and significance.

1. A premise for the sustainable development of the Chinese ancient civilization.

Of five thousand years' of ancient China, the first two thousand years saw a small scale of population, land development and economy. Starting from the Warring States Period, metal farming tools were adopted and practiced extensively which led to the boom of agricultural productivity.

Up to the Qing Dynasty, population had come to around 4 billion, which resulted in large scale of land reclamation and cultivation. This had exerted an enormous pressure on the ecological environment in order to support such a big population. Since then, the ecological environment had been rapidly deteriorating, especially in the north of China.

Though there were fluctuations in the developmental process of the Chinese ancient civilization, the relation among population, land development and technical levels had been relatively coordinated.

2. A foundation for the sustainable development of the ancient Chinese civilization.

The agricultural farming conditions in China have always been diversified due to its vast territory. Compared with the prosperous planting farming of the south, the north is covered by vast grasslands for livestock farming. In ancient times, the Great Wall was built by the Han people to fence the nomadic minorities to invade the south, which in effect became the boundary of planting and livestock breeding and farming. As a result, the planting and breeding were developed respectively in the designated regions.

Since early ancient times, the Chinese people had been utilizing manure and irrigation, taking Dujiangyan (都江堰) as an extraordinary example throughout the history until today. On the basis of spreading manure and scientific irrigating, intensive cultivation and protection of the land were carried out. Some western researchers applauded this farming model and called it "Peasant-gardener system".

In general, land resource had been conserved owing to scientific combination of planting and breeding, and scientific irrigating, spreading manure and intensive cultivation, which laid a solid foundation for the sustainable development of the ancient Chinese civilization.

3. Another significant reason: the shifting of civilization center.

Throughout the Chinese history, there was a gradual shifting of civilization center from the Yellow River Valley to Yangtze River Valley, especially after Qin and Han Dynasties. The main reason for this is the ecological changes caused by frequent wars and over-farming in the north. During the period of the Northern Song Dynasty, the population of the south had overturned the north in taking the lead to develop the national economy.

In short, with the south beginning to increasingly ship huge amounts of grain, textile northward as well as paying more taxes, the ecological pressure in the north had been relieved indirectly along with the shifting.

(Excerpted from "On the relation between sustainable development of Chinese ancient civilization and ecologic environment", *China Environmental Science*, Vol.16 No. 3, June 1996.)

Vocabulary

harmony between man and nature 天人合一（人与自然和谐共处）
sustainable development 可持续发展
premise 前提　　　foundation 基础　　　shifting 转移　　　dynasty 朝代

Reflection and Activities

1. Are there any other traditional Chinese thoughts that you are aware of being beneficial to the establishment of ecological civilization? Team up, make some investigations and see if you can come up with a mini-report on your intended topic.

Your report may consist of 3 parts:

A. Theme exploration.

B. Research design.

C. Findings and implications.

2. "Harmony between man and nature" is a household slogan among people. It affects every part of our life, can you name some examples (project/architecture/activity, etc.), and make a detailed presentation within your learning group or team?

第九节 《论中国古代文明的可持续发展与生态环境的关系》

导读

孔繁德出生于1945年4月。中国环境管理干部学院（现为河北环境工程学院）生态研究所所长，河北环境工程学院教授，秦皇岛市政协委员，致力于生态保护的教学与研究及秦皇岛环境保护与发展研究。

1982年，孔繁德教授以环保主义者的身份开始了自己的职业生涯。在从事生态保护研究的过程中，孔繁德教授发表论文30余篇，出版教材与专著10本，主持科研项目20余项。其中，《城市环境规划及方法指南》1995年获国家教委、国家环保局、全国环境教育优秀教材三等奖。《城市生态环境建设与保护规划》2001年获秦皇岛市委、市政府"五个一工程奖"。《中小城市环境规划的方法研究》获国家环保局科技进步三等奖。

孔繁德教授亲眼目睹了中国环保事业的实际启动和发展。直到2008年退休，他通过研究环境保护与秦皇岛蓬勃发展的经济之间的关系，为这项事业做出了巨大贡献。在他26年的职业生涯中，他向秦皇岛市政协提交了30多份提案，并在当地政协会议上发表了10多次演讲，对城市蓝图的设计产生了重大影响，包括在海港区西部建立一个新区到目前为止已成为现实。

孔繁德教授对城市环境保护研究的贡献，是中国生态文明繁荣的宝贵财富。

他的研究和思想主要集中在"天人合一"的概念上。

1. 在古代发展思想中，探讨"天人合一"思想具有重要意义，是中国哲学中备受关注和不断更新的话题。"天人合一"思想中蕴含的简单可持续发展理念至今仍闪耀着智慧之光。中国幅员辽阔，自古以来就是一个农业大国。在古代，生产力低下时，人类在自然面前极其脆弱。我国古代圣人从原始宗教信仰中对生命和自然的敬畏，农业文明时期人对自然的依赖出发，提出了"天人合一"的思想，强调人的行为应与自然规律相协调，道德理性应与自然理性相一致。

在现代高科技时代，人们利用科学技术积极开发自然资源。这严重破坏了自然的平衡。人类也面临着严重的生存危机。中国国家主席习近平深受中国传统文化影响，他从历史的角度探索中国生态文明建设的道路，继承了中国古代"天人合一"思想的生态智慧，形成了习近平的生态文明思想，指出了新时期生态环境保护的方向。

2. 中国传统儒家以人是自然的组成部分为主要思想之一，强调人与生态的统一。"天人合一"

是周易创造的，由孔子、孟子、荀子发展，最终由董仲舒创立，并由二程、朱子（朱熹）和王阳明拓充了内容。从那时起，这个想法就以一种简单直接的方式完成和表达了。

"天人合一"包括"三才"的和谐思想。从现代生态学的角度来看，这些观点激发了环境保护意识，体现了生态和谐的思想，展示了适度保护的准则，催生了环境保护的措施。然而，由于历史条件的限制，"天人合一"理论在采取实际措施和具体行动方面也有其自身的缺陷，与现代观念不相适应。

3. 思想的现代转换涉及两个方面：首先，现代社会继承了"天人合一"思想中的实践元素。我们的社会需要实现统一的思想，协调人与自然的关系，吸收仁爱的情感，关爱每一个生物。我们应该用生态平衡的原则来保护环境，遵守诚信伦理来提升自然资源的功能。第二，要处理好这一思想的缺陷，使之在新时期得到重新界定和完善。

节选

中国古代文明的可持续发展在世界上是罕见的，为后人留下了丰富的历史经验和智慧。中国古代文明的可持续发展有多种原因。其中，生态环境变化具有重要意义。

1. 中国古代文明可持续发展的前提。

在中国古代五千年中，前两千年人口、土地开发和经济规模较小。从战国时期开始，金属农具被广泛采用和使用，导致了农业生产力的繁荣。

到清代，人口已达40亿左右，导致大规模的土地开垦和耕作。为了养活这么多的人口，对生态环境造成了巨大的压力。从那时起，生态环境迅速恶化，特别是在中国北方。

虽然中国古代文明的发展过程有波动，但人口、土地开发和技术水平之间的关系相对协调。

2. 中国古代文明可持续发展的基础。

中国幅员辽阔，农业生产条件一直是多样化的。与南方繁荣的种植业相比，北方有广阔的畜牧业草原。在古代，长城是由汉族人修建的，用来围住游牧民族入侵南方，实际上成为种植业、畜牧业和农业的边界。因此，在指定区域分别开展种植和养殖。

自古以来，中国人就利用肥料和灌溉，占领都江堰在历史上直到今天都是一个非凡的例子。在施肥和科学灌溉的基础上，进行了集约化栽培和土地保护。一些西方研究人员称赞这种耕作模式，称之为"农民园丁制度"。

总体上，由于种植与养殖科学结合、科学灌溉、推广施肥和集约栽培，土地资源得到了节约，为中国古代文明的可持续发展奠定了坚实的基础。

3. 另一个重要原因：文明中心的转移。

纵观中国历史，尤其是秦汉以后，文明中心逐渐从黄河流域向长江流域转移。主要原因是北方频繁的战争和过度耕作造成的生态变化。北宋时期，南方的人口超过了北方，率先发展国民经济。

简言之，随着南方开始越来越多地将大量粮食、纺织品运往北方，并支付更多的税收，北方的生态压力随着人口的转移而间接缓解。

（节选自《中国环境科学》1996年6月第16卷第3期《论中国古代文明的可持续发展与生态环境的关系》）

思考题

1. 你是否意识到中国还有其他有利于生态文明建设的传统思想？以团队合作形式做一些调查，看看你是否能就你想要的主题提出一份报告。

报告可以由三部分组成：

A. 主题探索

B. 研究设计

C. 调查结果和影响

2. "天人合一"是人们家喻户晓的口号。它影响着我们生活的方方面面，你能举出一些例子（项目/架构/活动等），并在你的学习小组或团队中做一个详细的介绍吗？

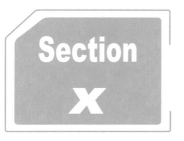

Section X
Sustainable Management of Water Resources

On May 21, 2016, the 2016 Annual Meeting of the Environmental Management Committee of the Chinese Society of Management Sciences and the unveiling ceremony of Hebei University of Environmental Engineering were held in Beidaihe. Qian Yi, academician of the Chinese Academy of Engineering, and Qu Geping, honorary president of Hebei Environmental Engineering College, jointly inaugurated the school.

Introduction

Qian Yi, born in 1935, is a professor of Tsinghua University and academician of Chinese Academy of Engineering. For decades, she has devoted herself to the research and development of new technologies for high-efficiency and low-consumption wastewater treatment suitable for our country's reality; she has carried out fruitful work on the bio-degradation's characteristics, treatment mechanism and technology. Qian Yi has presided over national scientific and technological research projects such as "Anaerobic Biological Treatment Technology of High-Concentration Organic Industrial Wastewater", "Urban Wastewater Stabilization Pond" and "High-Concentration Toxic Industrial Wastewater Treatment Technology and Equipment", and has won the second prize of National Science and Technology Progress Award, State Technological Invention Award, the first prize and the second prize of Scientific and Technology Progress in Education Ministry, the first prize of Natural Price for Natural Science given by Chinese Academy of Sciences, and won the highest prize for International Co-operation on Environment and Development. She has published 16 books and more than 360 papers.

Qian Yi has been engaged in environmental engineering teaching and research for more than 50 years, and has achieved remarkable results in the fields of industrial wastewater treatment and urban wastewater purification. (Review from "Looking for the Most Beautiful Teacher")

Since Qian Yi devoted herself to education for half a century, she has influenced and cultivated several generations in the field of environmental science and engineering in China, and created a large number of academic backbones. As a pioneer of quality education for environmental protection and sustainable development in Chinese institution of higher learning, she is the first person in Tsinghua University to advocate "building a 'green university'" and has made important contributions to the development of environmental science and engineering education in China. (Review from Tsinghua University Outstanding Contribution Award)

In 2016, on Qianyi's 80th birthday, her students initiated and donated money to establish the "Qianyi Environmental Education Fund", and based on this fund, the "Qianyi Environmental Award" was launched. Its purpose is to inspire outstanding students in China to actively practice environmental public welfare and carry out innovative research in a down-to-earth manner.

Excerpts

Water is a very valuable resource, sustaining human life, production processes, and ecosystems; thus, particular attention should be paid to the management of water resources. China is now facing a serious water crisis, including water shortages, flooding, and water pollution, due to both natural and artificial causes. This water crisis has threatened human health and economic development.

Natural water resources in China have two main characteristics. First, they are insufficient: Water availability in China is only 1780 m^3 per capita, which is one quarter that of the global value and very close to the universally accepted figure for countries lacking water resources. Second, they are unevenly distributed: Although there are rich water resources in South China, there is a lack of water in North China. North China possesses 47% of the population and 65% of the farmland, but only 17% of China's water resources. In addition, rainfall is unevenly distributed over the four seasons in China.

Water pollution is caused by human activities. Industry, agriculture, and the activities of daily life all discharge wastewater and create water pollution. The seven major water systems in China all contain pollution to different degrees, including organic pollution, nutrient pollution, heavy metal pollution, persistent organic pollutants (POPs), and so forth. It is estimated that more than 100 million urban residents and 320 million residents in rural areas lack safe drinking water. Many accidents related to water pollution have occurred in China in recent years, seriously threatening people's health and lives.

Flooding and water-logging are sudden events. However, most of the responsibility for these issues lies with sewer systems that are not well-designed, constructed, and/or operated. For example, some cities in North China have had flood-related accidents that were mainly caused by problems with urban sewer systems.

All these facts make it clear that close attention should be paid to the management of water resources in China in order to prevent the harmful effects of water crises. The targets of sustainable water resource management should include: providing sufficient water resources for economic development and the population's daily lives; protecting a clean water environment; ensuring safe drinking water; and preventing disasters caused by flooding and water-logging.

To ensure that the above targets are achieved, the sustainable management of water resources in China should be based on these strategies:

1. Making water conservation the main priority, and controlling water demand;

2. Controlling water pollution rigorously and reducing the pollution created by sources;

3. Using wastewater as water, energy, and fertilizer resources;

4. Preventing flooding and water-logging disasters, while using rain water as a water resource.

(Excerpted from "Sustainable Management of Water Resources", *Engineering* 2 (2016) 23-25, Qian Yi, School of Environment, Tsinghua University, Beijing 100084, China)

Vocabulary

water availability 水资源可利用量

persistent organic pollutants (POPs) 持久性有机污染物

sewer systems 排水系统

Reflection and Activities

1. The sustainable management of water resources in China emphasizes on making water conservation the main priority, and controlling water demand. What kind of tips do you have for saving water in daily life?

2. The "World Water Day and China Water Week" will be held soon. To arouse public water conservation concept, a strong publicity atmosphere should be created by making posters and distributing relevant materials. Please design and create English posters on this theme.

《水资源的可持续管理》

导读

钱易，1935年生，清华大学教授，中国工程院院士。数十年来致力于研究开发适合我国国情的高效、低耗废水处理新技术；对难降解有机物生物降解特性、处理机理及技术进行了卓有成效的工作。曾主持"高浓度有机工业废水的厌氧生物处理技术"、"城市废水稳定塘"及"高浓度有毒工业废水处理技术及设备"等国家科技攻关课题，并获国家科技进步二等奖一次、三等奖两次，国家技术发明奖一次，国家教委科技进步一等奖两次、二等奖两次、中国科学院自然科学一等奖一次，获得中国环境与发展国际合作最高奖。完成著作16部，发表论文360余篇。

钱易从事环境工程教学与科研50余年，在工业废水处理与城市废水净化等领域取得了令国际环境工程界瞩目的成果。(《寻找最美教师》评)

钱易投身教育事业半个世纪以来，影响和培养了中国环境科学与工程领域的几代人，造就了一大批学术骨干。作为中国高等院校环境保护与可持续发展素质教育的先行者，她是清华大学倡导"建设'绿色大学'"的第一人，为中国环境科学与工程教育事业的发展做出了重要贡献。(清华大学突出贡献奖评)

2016年，在钱易80寿辰时，她的学生们发起并捐资成立了"钱易环境教育基金"，依据该基金设立了"钱易环境奖"，宗旨是激励中国积极践行环境公益、脚踏实地开展创新研究的优秀学生。

节选

水是生命之源、生产之本、生态之基。因此，应该特别注意水资源管理。由于天然和人为的原因，中国正面临着严重的水资源危机，包括：水资源短缺、洪涝灾害以及水体污染。水资源危机已经威胁到了人类的健康和经济的发展。

中国天然水资源主要有两个特点。一是水资源不足：水资源可利用量约为人均1780 m³，仅为全球水资源可利用量的四分之一，与国际上公认的用水紧张国家(水资源可利用量小于人均1700 m³)非常接近。二是水资源地域分布不均：呈现出南多北少的局面。北部地区居住了全国47%的人口，拥有全国65%的耕地面积，但水资源拥有量却只有全国水资源量的17%。此外，中国的降水还具有四季分布不均的特点。

水体污染主要由人为原因造成。工业、农业和人们的日常生活都会排放含有各种污染物的污水、废水，造成水体污染。中国的七大水系都已经被不同程度污染，包括有机污染、富营养污染、重金属污染和持久性有机物污染等。据估计中国至少有1亿城市人口、3.2亿农村人口缺乏安全的饮用水。近几年发生了许多与水体污染有关的事故，严重威胁着人们的健康和生命。

洪涝灾害一般都是突发事件。然而，问题产生的大部分原因在于未对排水系统进行精心设计、建造以及运行维护。例如，由于城市排水系统的问题，我国北方的一些城市会遭受洪灾。

为防止水资源危机的有害影响，中国应密切重视可持续水资源管理。可持续水资源管理的目标应该包括：为经济发展及人们的日常生活提供充足的水资源，保护清洁的水环境，保障安全的饮用水，防止洪涝灾害。

为确保上述目标的实现，中国的可持续水资源管理应采取以下策略：

1. 节约用水，控制需求；
2. 严控污染，抓好源头；
3. 努力实现废水的资源化、能源化；
4. 变洪水为资源，防止洪涝灾害。

（节选自钱易在2016国际城市低影响开发学术大会的报告《水与可持续发展》）

思考题

1. 中国水资源的可持续管理强调以节约用水为主，控制用水需求。日常生活中你有哪些节水小窍门？

2. 世界水日和中国水周即将举行。唤起公众节约用水的观念，应通过制作海报和发放相关材料，营造浓厚的宣传氛围。请为此主题设计和制作英文海报。

Chapter II

Xi Jinping Thought on Eco-civilization

In May 2018, the CPC Central Committee held a national conference on ecological and environmental protection. At the meeting, President Xi Jinping delivered an important speech and put forward six important principles that we must stick to in order to promote the construction of eco-civilization in the new era. The six principles are harmony between man and nature; lucid waters and lush mountains are invaluable assets; a good eco-environment is the most inclusive form of public wellbeing; mountains, rivers, forests, farmlands, lakes and grasslands are a community of life; protect the ecological environment with the strictest system and the strictest rule of law and work together to promote a global eco-civilization. Summarized from long-term practice, the six principles considering the development of all mankind are the very essence of Xi Jinping thought on eco-civilization.

In this chapter, the connotation, significance and specific practical methods of the six principles will be introduced one by one in the following six sections. It aims to help students have an in-depth understanding of the problems existing in China's environmental protection and how Xi Jinping thought on eco-civilization points out the direction for sustainable development.

第二章

习近平生态文明思想

2018年5月,党中央召开全国生态环境保护大会。会上,习近平总书记发表重要讲话,对推进新时代生态文明建设提出必须遵循的六项重要原则,即坚持人与自然和谐共生、绿水青山就是金山银山、良好生态环境是最普惠的民生福祉、山水林田湖草是生命共同体、用最严格制度和最严密法治保护生态环境及共谋全球生态文明建设。站在人类发展的高度,从长期实践中总结形成的"六大原则",是习近平生态文明思想的精髓。

本章的六节将逐一介绍这六大原则的内涵、重要意义及其具体实践方法,旨在帮助学生深入了解我国环境保护方面存在的问题以及习近平生态文明思想是如何为可持续发展指明了方向。

Harmony between Man and Nature

Introduction

Xi Jinping's thought on ecological civilization is an interpretation of the three-in-one relationship of human, nature and society, and it is embodied in eight dimensions, including view of history, view of nature, view of development, view of people's livelihood, view of system, view of rule of law, view of co-governance, and view of globalization.

It has dual value of the times at the theoretical level and the practical level, for it enriches the Marxist theoretical connotation of the relationship between man and nature, carries forward the simple ecological wisdom in China's traditional culture, and deepens the cognitive relationship between China's social development and environmental protection. At the same time, it is also conducive to guiding the overall promotion of China's "five-in-one" construction and contributing Chinese wisdom and Chinese solutions to global ecological governance.

Since the beginning of the 21st century, due to the industrialization process, the ecological environment crisis has intensified all over the world, and the contradiction between economic development and environmental protection is widespread. Dealing with the relationship between the two has become a hot topic in today's world.

Since the reform and opening up, with the continuous improvement of China's openness and the vigorous development of the socialist market economy with Chinese characteristics, China has made remarkable development achievements. At the same time, the previous development mode based on efficiency and speed has also greatly damaged China's ecological environment. The report of the 19th National Congress of the Communist Party of China put forward the principle of "consisting on the harmonious coexistence of man and nature" and becoming an important follower to lead the party and the people in building the cause of socialism with Chinese characteristics. Xi Jinping's harmonious symbiosis between man and nature is an important part and important content of his new era of ecological civilization.

Xi Jinping has a profound foundation for the formation of the theory of harmonious symbiosis between man and nature. He insisted on Marx and Engels' thoughts on the harmonious relationship between man and nature, inherited the thoughts of the leaders of the Communist Party of China on the relationship between man and nature and absorbed the thoughts of "the combination of heaven and man" of Confucianism and Taoism in ancient Chinese traditional culture. He puts forward the theory of harmonious symbiosis between man and nature in practice.

Xi Jinping's theory of harmonious symbiosis between man and nature has rich connotations. The main contents of this thought include the following: The concept of respecting nature conforming to nature, and protecting nature. The green development concept of lucid waters and lush mountains are invaluable assets. The theory of protecting the environment means protecting productivity, improving the environment means developing productivity and building a clean and beautiful world.

On how to achieve a harmonious symbiosis between man and nature, Xi Jinping clearly pointed out the practice path. Persist in the basic national policy of saving resources and protecting the environment. With the most rigorous system, the strictest rule of law to protect the ecological environment. Form a green way of development and lifestyle. Xi Jinping's theory of harmonious symbiosis between man and nature has distinct characteristics. It reflects his profound feelings of the people, distinctive characteristics of the times, strong problem awareness and broad international vision.

The theory of harmonious symbiosis between man and nature is of great significance. It advances the Marxist idea of harmony between man and nature to a new height. It is conducive to the resolution of the main contradictions in China's new era society, is conducive to the realization of the goal of beautiful China, and is conducive to the sustainable development of all mankind.

Excerpts

Sixth, let us build a China-Africa community with a shared future that promotes harmony between man and nature. The Earth is the only place which we mankind call home. China will work with Africa to pursue green, low-carbon, circular and sustainable development and protect our lush mountains and lucid waters and all living beings on our planet. We will strengthen exchange and cooperation with Africa on climate change, clean energy, prevention and control of desertification and soil erosion, protection of wildlife and other areas of ecological and environmental preservation. Together, we could make China and Africa beautiful places for people to live in harmony with nature.

(Excerpted from Chinese President Xi Jinping's speech at opening ceremony of 2018 FOCAC Beijing Summit)

Vocabulary

ecological civilization 生态文明

harmonious symbiosis 和谐共生

low-carbon 低碳

Chinese characteristics 中国特色

Marxist 马克思主义

circular 循环的

Reflection and Activities

1. Xi Jinping's harmonious symbiosis between man and nature is embodied in nearly all aspects of our life, whether in life or at work. It is highly suggested that you find out one example to tell your teacher or friends about how the concept is embodied in actual life.

2. Watch the video on "How to achieve sustainable development in China" from CGTN and answer the following questions:

(1) What is Chinese ecological civilization?

(2) What has China been doing to control environmental pollution?

(3) What are the effective outcomes of Chinese pollution control?

第一节 人与自然和谐共生

导读

习近平的生态文明思想是对人、自然、社会三者一体关系的解读，它体现在历史观、自然观、发展观、民生观、制度观、法治观、共治观、全球化观八个维度上。

它在理论层面和实践层面具有双重的时代价值，丰富了马克思主义关于人与自然关系的理论内涵，弘扬了中国传统文化中朴素的生态智慧，深化了中国社会发展与环境保护的认识关系，也有利于指导中国全面推进"五位一体"建设，为全球生态治理贡献中国智慧和中国解决方案。

进入 21 世纪以来，由于工业化进程，全球生态环境危机加剧，经济发展与环境保护的矛盾普遍存在。如何处理两者之间的关系已成为当今世界的热门话题。

改革开放以来，随着中国对外开放程度的不断提高和中国特色社会主义市场经济的蓬勃发展，中国取得了显著的发展成就。与此同时，以往基于效率和速度的发展模式也对中国的生态环境造成了极大的破坏。中国共产党第十九次全国代表大会报告提出了"人与自然和谐共生"的原则，成为引领党和人民建设中国特色社会主义事业的重要遵循。习近平的人与自然和谐共生是其生态文明新时代的重要组成部分和重要内容。

习近平对人与自然和谐共生理论的形成有着深厚的基础。他坚持马克思恩格斯关于人与自然和谐关系的思想，继承了中国共产党领导人关于人与自然关系的思想，吸收了中国古代传统文化中儒、道"天人合一"的思想。他在实践中提出了人与自然和谐共生的理论。

习近平的人与自然和谐共生理论具有丰富的内涵。这一思想的主要内容包括：尊重自然、顺应自然、保护自然的观念。树立和践行绿水青山就是金山银山的理念。保护环境就是保护生产力，改善环境就是发展生产力，建设清洁美好的世界。

在如何实现人与自然和谐共生的过程中，习近平明确指出了实践路径。坚持节约资源、保护环境的基本国策。用最严格的制度、最严密的法治来保护生态环境。形成绿色的发展方式和生活方式。

习近平的人与自然和谐共生理论具有鲜明的特点。它反映了他深厚的民情、鲜明的时代特征、强烈的问题意识和广阔的国际视野。

人与自然和谐共生理论具有重要意义。它把马克思主义人与自然和谐的思想推向了一个新的高度。有利于解决新时期中国社会的主要矛盾，有利于实现美丽中国的目标，有利于全人类的可持续发展。

节选

　　第六，携手打造和谐共生的中非命运共同体。地球是人类唯一的家园。中国愿同非洲一道，倡导绿色、低碳、循环、可持续的发展方式，共同保护青山绿水和万物生灵。中国愿同非洲加强在对应气候变化、应用清洁能源、防控沙漠化和水土流失、保护野生动植物等生态环保领域交流合作，让中国和非洲都成为人与自然和睦相处的友好家园。

（节选自中国国家主席习近平在2018年中非合作论坛北京峰会开幕式致辞）

思考题

　　1. 习近平的人与自然和谐共生体现在我们生活的方方面面，无论是在生活中还是在工作中。请你找出一个例子，告诉你的老师或朋友这个概念在实际生活中是如何体现的。

　　2. 观看选自CGTN的视频"中国如何实现可持续发展"，回答以下问题：

（1）什么是中国的生态文明？

（2）中国在控制环境污染方面做了哪些努力？

（3）中国污染控制的显著结果是什么？

Section ii Lucid Waters and Lush Mountains Are Invaluable Assets

Introduction

1. General introduction

In August 2005, Xi Jinping, former Secretary of the Zhejiang Provincial Party Committee of the Communist Party of China, first put forward the scientific thesis that "Lucid waters and lush mountains are invaluable assets" during a survey in Anji County, Huzhou, Zhejiang Province. In March 2015, the "Opinions on Accelerating the Construction of Ecological Civilization" officially wrote "Lucid waters and lush mountains are invaluable assets" into the central files, and the concept of "Lucid waters and lush mountains are invaluable assets" (referred to as "Two Mountains" concept). It has risen to become the basic strategy and important national policy of governing the country. On May 18–19, 2018, the National Conference on Ecological Environmental Protection was held, and "Lucid waters and lush mountains are invaluable assets" became an important connotation of Xi Jinping Thought on Ecological Civilization.

"Lucid waters and lush mountains are invaluable assets" shows the green development concept of Xi Jinping's ecological civilization thought. Green mountains are not opposite to invaluable assets. To protect the ecological environment is to protect the productive forces, and to improve the ecological environment is to develop the productive forces. President Xi Jinping believes that the construction of ecological civilization and economic growth are both contradictory and unified. He vividly compared the relationship between the two to two mountains, gold mountains and silver mountains. But we also pursue valuable assets. He emphasizes that we should pursue harmony between man and nature and harmony between economy and society.

"Lucid waters and lush mountains are invaluable assets" is a scientific concept, formed by President Xi Jinping and established in the ecological practice of contemporary China. It is on the basis of inheriting and developing the Marxist ecological concept, drawing on the ecological wisdom of China's excellent traditional culture, and drawing on the experience of western countries in managing ecological and environmental problems. It is the core essence of Xi Jinping's Ecological Civilization Thought. Its essence is to adhere to the mutual benefit and win-win situation of ecological environmental protection and economic development so as to build a beautiful China.

2. Development Path

The key to implementing the concept of "Lucid waters and lush mountains are invaluable assets" lies in taking the five major development concepts of "innovation, coordination, green, openness and sharing" as the guide, taking the economic development path of industrial ecology and ecological industrialization, and promoting the establishment and improvement of the ecological civilization system.

(1) Industrial Ecology: In every industry and every region, it is necessary to combine ecology with industry, organize and manufacture in accordance with the way of protecting resources, conserving resources, and utilizing resources, and give ecological characteristics to the first, second, and third industries, so that mountains of gold and silver will not only have development value but also have high-quality living value because the lucid waters and lush mountains will be realized, and the return of mountains of gold and silver to green mountains at a higher level will be realized.

(2) Ecological Industrialization: Ecological industrialization is to realize the transformation of

ecological resources of green mountains into industrial resources of mountains of gold and silver, so that ecology can be developed and protected through industry. It is necessary to develop, utilize, and protect lucid waters and lush mountains with the development of ecological agriculture, ecological industry, and ecological service industry, so as to realize the dialectical unity of the two development concepts.

Appendix: Typical models and main features of the "Two Mountains" concept practice

The "Two Mountains" concept practices a typical model	Key features	Typical area(s)
Pollution control type	Solve the environmental pollution leftover from history in the process of economic and social development and the environmental pollution of the current production and living process, and improve the quality of the environment	Comprehensive improvement of rural human settlement environment in Anji County, Zhejiang Province; Comprehensive management of Luanchuan Mining in Henan
Ecological protection and restoration type	Strengthen artificial ecological restoration; make up for regional ecological shortcomings or restore the problems of ecological destruction brought about by exploitation of resource and energy; consolidate the ecological background	Saihanba Forest Farm in Hebei Province, Changting, Fujian Province and other desertification and soil erosion areas; resource-depleted cities such as Zixing, Hunan Province
Traditional industry adjustment type	Eliminate small enterprises, scattered and disorderly or highly polluted; strictly enforce environmental protection access thresholds, and implement a one-vote veto system for environmental protection of heavy pollution enterprises to optimize the industrial structure	Zhejiang Anji, Hunan Zixing and other areas were once dominated by high-pollution and high-energy consumption industries.
Energy-clean type	Develop and use clean energy such as wind power, hydropower, and photovoltaic power generation, and promote energy conservation and consumption reduction	Ninghai County, Zhejiang Province, built the country's largest marine aquaculture "fishing-light complementary" photovoltaic power generation project.
Green and intelligent manufacturing type	Guided by the green concept, we will comprehensively consider environmental impact and resource efficiency to develop modern or intelligent manufacturing industries, and promote the transformation of the industry to a green cycle and low-carbon one	The industrial output of new membrane materials, electronic information, machinery manufacturing "2 + 1" accounted for more than 30% in Sihong, Suqian, Jiangsu Province.

The "Two Mountains" concept practices a typical model	Key features	Typical area(s)
High-end industrial development type	Based on the advantages of a good ecological environment, we will attract high value-added and low-pollution industries, and give rise to new economies, new formats and new models.	Mengyin, Shandong Province, with annual 4.8 billion yuan turnover of e-commerce industry, became China e-commerce demonstration county.
Efficient ecological agriculture type	Give priority to the advantages of local agricultural products, medicinal herbs, forest products, seafood and livestock products, and form a brand influence and scale effect	Tengchong, Yunnan Province has been working on building the best planting (breeding) area of Chinese herbal medicine among counties.
Efficient ecological breeding type		
Efficient eco-industrial type	Based on the advantages of eco-agriculture and eco-breeding, a whole industrial chain system of breeding, processing, sales, scientific research, innovation and health care is formed	Chishui City, Guizhou Province, has been building an ecological industrial cluster with the gross output value of 50 billion
Global eco-tourism	Support infrastructure; develop boutique homestays; create boutique tourism products; create a global tourism pattern	Red tourism in Jinggangshan City, Jiangxi Province, accounts for more than 50% of GDP, and is a national global tourism demonstration city
Ecological and cultural industry type	Give full play to the attractiveness of red culture, historical culture and folk culture, and promote the development of cultural industries	
Tertiary integration development	With the help of new models such as "Internet +", "Tourism +" and "Ecology +", we will promote the deep integration of related industries such as agricultural and commercial cultural tourism	Huzhou, Zhejiang Province, has been building a three-industry integration base of bamboo-tea-mulberry-lake sheep-red plum

(Excerpted from "Practice Mode and Path of 'Lucid Waters and Lush Mountains are Invaluable Assets'", Dong Zhanfeng, 【DOI】10.16868/j.cnki.1674-6252.2020.05.011)

Excerpts

1. Lucid waters and lush mountains are invaluable assets and improving ecological environment is increasing productivity.

2. We should pursue harmony between man and nature.

3. We should maintain the overall balance of Earth's ecology, so that the starry sky, lush mountains

and floral fragrance will be retained for our future generations who enjoy material prosperity at the same time.

4. We should pursue the prosperity based on green development.

5. Only with concerted efforts can we effectively deal with global environmental issues such as climate change, marine pollution and biological protection, and achieve the United Nations 2030 Agenda for Sustainable Development goals.

6. We should foster a passion for nature-caring lifestyle.

7. We should embrace simple, moderate, green and low-carbon ways of life, and reject extravagance and waste of resources, making the idea of ecological and environmental conservation the mainstream culture in society.

8. We should pursue a scientific spirit in ecological governance.

9. Ecological governance should follow the law of nature and make scientific planning, while taking into consideration local conditions and a holistic approach to create a diversified ecological system of co-existence.

(Excerpted from a speech delivered by Chinese President Xi Jinping at the opening ceremony of the International Horticultural Exhibition 2019, Beijing)

Vocabulary

lucid 清澈的
invaluable 极宝贵的
ecological industrialization 生态产业化
lush 茂盛的
industrial ecology 产业生态化

Reflection and Activities

1. A herbal medicine planting project will offer about 5,000 job positions and motivate local economic development. However, some changes, such as plant cutting for the supporting modern facilities, occupation of large numbers of farmland might take place as well.

(1) Suppose that you are a mayor, will you carry out the project?

(2) There are often conflicts between economic development and environmental protection. How should we balance the two and ensure sustainable development?

2. "Go Green Week" is just a beginning to a greener life and finally a greener world! Are you inspired

to have your own "Go Green Week"?

    3. Plan a "Go Green Week" for yourself and complete the following work sheet.

My "Go Green Week"	
Monday	1. Eat seasonal and local food 2. To improve my health 3. Reduce gas emissions caused by transport
Tuesday	
Wednesday	
Thursday	
Friday	

第二节 绿水青山就是金山银山

导读

1. 概述

2005年8月,时任中共浙江省委书记的习近平,在浙江湖州安吉县调研时首次提出了"绿水青山就是金山银山"的科学论断。2015年3月,《关于加快推进生态文明建设的意见》正式把"坚持绿水青山就是金山银山"写进中央文件,"绿水青山就是金山银山"理念(简称"两山"理论)上升为治国理政的基本方略和重要国策。2018年5月18—19日,全国生态环境保护大会召开,"绿水青山就是金山银山"成为习近平生态文明思想重要内涵。

绿水青山和金山银山不是对立的,保护生态环境就是保护生产力,改善生态环境就是发展生产力。习近平总书记认为,生态文明建设与经济增长既存在矛盾,又可以统一。他把二者的关系形象地比作两座山,"既要绿水青山,也要金山银山",强调我们追求人与自然的和谐,经济与社会的和谐。

"绿水青山就是金山银山"是习近平总书记立足当代中国生态实践,在传承发展马克思主义生态观、汲取中华优秀传统文化中国的生态智慧、借鉴西方国家治理生态环境问题的经验基础上形成的科学理念,是习近平生态文明思想的核心要义。其实质是坚持生态环保和经济发展互利共赢,建设美丽中国。

2. 发展路径

贯彻"绿水青山就是金山银山"理念的关键在于,以"创新、协调、绿色、开放、共享"的五大发展理念为指导,走产业生态化和生态产业化的经济发展路径,推动生态文明体系建立和完善。

① 产业生态化。产业生态化就是用绿水青山的生态价值提升金山银山的产业形态,在经济发展中保护和修复生态,实现人的发展与自然存在的和谐共处。因此,必须按照生态化的原则优化和改造原有生产方式。在每个行业和每个地区中,都要把生态和产业结合起来,按照保护资源、节约资源、利用资源的方式组织生产,给一二三产业赋予生态化的特色,让金山银山因为绿水青山不仅具备发展价值而且有高质量的生活价值,实现金山银山在更高层次上向绿水青山的回归。

② 生态产业化。生态产业化就是实现绿水青山的生态资源向金山银山的产业资源转变,让生态通过产业的方式得到开发和保护。要以生态农业、生态工业和生态服务业的生态型产业发展去开发、利用和保护绿水青山,实现两种发展理念的辩证统一。

附表："两山"理念实践典型模式及主要特征

"两山"理念实践典型模式	主要特征	典型地区
污染治理型	解决经济社会发展进程中的历史遗留环境污染和当下生产生活过程的环境污染，改善提升环境质量	浙江安吉县农村人居环境综合整治；河南栾川矿业综合治理
生态保护修复型	加强人工生态修复，补齐区域生态短板抑或是修复资源能源开采等带来的生态破坏问题，夯实生态本底	河北塞罕坝林场、福建长汀等荒漠化、水土流失地区；湖南资兴等资源枯竭型城市
传统产业调整型	淘汰低小散污企业，严格环保准入门槛，实行重污染企业环保一票否决制等，优化产业结构	浙江安吉、湖南资兴等曾以高污染高能耗产业为主地区
能源清洁化型	开发使用风电、水电、光伏发电等清洁能源，推动节能降耗	浙江宁海县建设全国最大的海水养殖"渔光互补"光伏发电项目
绿色制/智造型	以绿色理念为指导，综合考虑环境影响、资源效率发展现代制/智造产业，推动产业向绿色循环低碳方向转变	江苏泗洪新型膜材料、电子信息、机械制造"2+1"产业产值占比达30%以上
高端产业发展型	基于良好生态环境优势，吸引高附加值低污染产业，催生新经济、新业态和新模式	山东蒙阴电子商务产业年交易额48亿元，成为中国电子商务示范县
高效生态农业型	注重发挥当地特色农产品、药材、林产品、海产品、畜牧产品优势，形成品牌效应和规模效应	云南腾冲打造全国最优质的中草药种植（养殖）区
高效生态养殖型		
高效生态工业型	立足于生态农业、生态养殖等优势基础上，形成种养、加工、销售、科研、创新、康养的全产业链体系	贵州赤水市打造五百亿生态工业集群
全域生态旅游型	配套基础设施、开发精品民宿，打造精品旅游产品，打造全域旅游格局	江西井冈山市红色旅游业占GDP的比重超过50%，是国家全域旅游示范市
生态文化产业型	发挥红色文化、历史文化、民俗文化的吸引力，推动文化产业发展	
三产融合发展型	借力"互联网+""旅游+""生态+"等新模式，促进农商文旅等相关产业深度融合	浙江湖州打造竹—茶—桑—湖羊—红梅等三产融合基地

（节选自"'绿水青山就是金山银山'理念实践模式与路径探析"，董战峰，【DOI】10.16868/j.cnki.1674-6252.2020.05.011）

节选

1. 绿水青山就是金山银山，改善生态环境就是发展生产力。
2. 我们应该追求人与自然和谐。

3. 我们要维持地球生态整体平衡，让子孙后代既能享有丰富的物质财富，又能遥望星空、看见青山、闻到花香。

4. 我们应该追求绿色发展繁荣。

5. 唯有携手合作，我们才能有效应对气候变化、海洋污染、生物保护等全球性环境问题，实现联合国2030年可持续发展目标。

6. 我们应该追求热爱自然情怀。

7. 我们要倡导简约适度、绿色低碳的生活方式，拒绝奢华和浪费，形成文明健康的生活风尚。

8. 我们应该追求科学治理精神。

9. 生态治理必须遵循规律，科学规划，因地制宜，统筹兼顾，打造多元共生的生态系统。

（节选自2019年4月28日，国家主席习近平在北京延庆出席中国北京世界园艺博览会开幕式，并发表题为《共谋绿色生活，共建美丽家园》的重要讲话）

思考题

1. 一个草药种植项目将提供约5000个工作岗位，并激励当地经济发展。但是，例如为了支撑现代设施的植物砍伐，大量占用农田的情况也可能发生。

（1）假设你是市长，你执行该项目吗？

（2）经济发展与环境保护之间经常存在冲突。我们应该如何平衡两者并确保可持续发展？

2. 为自己计划一个"绿色"的一周，并完成以下工作表。

我的"绿色周"	
星期一	1. 吃时令和当地美食 2. 增强体质 3. 减少交通出行造成的废气排放
星期二	
星期三	
星期四	
星期五	

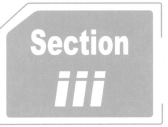

Section iii

A Good Eco-environment is the Most Inclusive Form of Public Wellbeing

Introduction

Over the past forty years of reform and opening up, China's economic development has made historic achievements, which is worthy of our pride. However, it must be pointed out that the ecological and environmental problems accumulated by the rapid development of more than 40 years will enter a high-intensity and frequent stage. The high incidence of various environmental pollution has become a disaster for people's livelihood. Seriously polluted weather, black and smelly water bodies, garbage besieged cities and varying degrees of rural environmental problems are still the suffering of the people. People are changing from "yearning for food and clothing" to "yearning for environmental protection", from "seeking survival" to "seeking ecology". People's needs are diverse, multi-level and multifaceted. They look forward to enjoying a better ecological environment. Chinese urban residents' consideration of livability has gradually surpassed the traditional happiness factors in our concept, such as per capita GDP, economic level and educational resources. The status of ecological environment in people's life happiness index is becoming more and more prominent. The destruction and pollution of ecological environment do not only affect the sustainable development of economy and society, but also affect people's health, which has become a prominent livelihood problem.

During his visit to Hainan Province in April 2013, President Xi Jinping first proposed that a good eco-environment is the most inclusive form of public wellbeing and the most equitable public goods. Green water and green mountains are an important part of people's happy life, and money can't replace them. Adhering to a good ecological environment is the most inclusive people's wellbeing and one of the six principles that must be followed in the construction of ecological civilization. It does not only reflect the people's position and feelings of General Secretary Xi's thought of ecological civilization, but also reveals the starting point, foothold and destination of the construction of ecological civilization.

Developing the economy is for the sake of the people's livelihood, and protecting the ecological environment is also for the sake of the people's livelihood. We should not only create more material and spiritual wealth to meet the people's growing needs for a better life, but also provide more high-quality ecological products to meet the people's growing needs for a beautiful ecological environment. This principle implements the people-centered concept. We should adhere to the principle of benefiting and serving the people by ecology, focus on solving prominent environmental problems that damage people's health, accelerate the improvement of the quality of the ecological environment, provide more high-quality ecological products, strive to achieve social fairness and justice, and constantly meet the people's growing needs for a beautiful ecological environment.

The practice of this principle needs to rely on the people. We will continue to implement air pollution prevention and control actions and win the battle to protect the blue sky. We will accelerate the prevention and control of water pollution and implement comprehensive treatment of the basin environment and coastal waters. We will strengthen soil pollution control and remediation, strengthen the prevention and control of agricultural non-point source pollution, and carry out actions to improve the rural living environment, strengthen the disposal of solid waste and garbage. We should strengthen publicity and education on ecological civilization, enhance the people's awareness of conservation, environmental protection and ecology, cultivate ecological ethics and codes of conduct. It is also indispensable to carry out green actions for the whole people, create a good atmosphere of caring for the ecological environment, and mobilize the whole society to take practical actions to reduce energy and resource consumption and pollution emissions, so as to contribute to ecological and environmental protection.

Excerpts

A good eco-environment is the most inclusive form of public wellbeing. We should choose to do the things that win the approval of the people, and avoid doing things that they oppose. A good environment is part of the public's wellbeing, green mountains and blue skies bring delight and happiness to the people. Economic development contributes to improving people's wellbeing, so does eco-environmental protection. We should create more material and cultural wealth to meet the people's growing expectation for a better a life, and effectively preserve the ecosystems to meet their growing expectation for a beautiful environment. We must pursue environmental benefits for the people with the emphasis on solving prominent problems that threaten their health. We must accelerate this work, provide more quality eco-products, and realize social fairness and justice.

Eco-environmental progress is a common cause that requires the participation and contribution of the general public. Its results will also be shared by the people. To build a beautiful China, we should transform our efforts into conscientious action on the part of all. Each and every individual is a protector, builder and beneficiary, and none should be a bystander, an outsider or a critic. No one should remain aloof and pay only lip service. We must enhance people's awareness of resource conservation, environmental protection, and a healthy ecosystem. We need to cultivate eco-friendly ethics and codes of conduct, launch nationwide green environment campaigns, and encourage the whole of society to contribute to environmental protection by reducing pollution and consumption of energy and other resources.

(Excerpted from the speech delivered by Chinese President Xi Jinping at the National Conference on Eco-environmental Protection on May 18, 2018)

Vocabulary

livelihood 民生

prominent 突出的；显著的

codes of conduct 行为准则

wellbeing 健康；福祉

beneficiary 受益者

Reflection and Activities

1. Read the guided reading part and underline the sentence which can indicate the special significance of this principle. Talk about your understanding about the statement.

2. The group of 4 or 5 investigates what has changed in the past three years in your hometown or the city you are living now, such as smog, drinking water safety, food safety, garbage siege, etc. Write an investigation report or take pictures, and draw a conclusion.

第三节 良好的生态环境是最普惠的民生福祉

导读

改革开放 40 年来，中国经济发展取得了历史性成就，这是值得我们骄傲的。但必须指出，40多年的快速发展所积累的生态环境问题必将进入高强度、频繁爆发的阶段。各种环境污染的高发已经成为民生的灾难。严重污染的天气、黑臭的水体、垃圾围城和不同程度的农村环境问题仍然是人民的苦难。人们从"向往衣食"到"向往环保"，从"求生存"到"求生态"。人民的需求是多样的、多层次的和多方面的，他们期待着享受一个更美好的生态环境。中国城市居民对宜居性的考虑已经慢慢超越了我们概念中的传统幸福因素，如人均 GDP、经济水平和教育资源。生态环境在人民生活幸福指数中的地位不断凸显。生态环境的破坏和污染不仅影响经济社会的可持续发展，而且影响人们的健康已成为突出的民生问题。

在 2013 年 4 月访问海南省期间，习近平主席首次提出，良好的生态环境是最普惠的民生福祉和最公平的公共产品。绿水青山是人们幸福生活的重要组成部分，金钱无法替代。坚持良好生态环境是最普惠的民生福祉，是生态文明建设必须遵循的六项原则之一，体现了习总书记生态文明思想的人民立场、人民情怀，也揭示了进行生态文明建设的出发点、立足点和归宿。

发展经济是为了民生，保护生态环境同样也是为了民生。既要创造更多的物质财富和精神财富以满足人民日益增长的美好生活需要，也要提供更多优质生态产品以满足人民日益增长的优美生态环境需要。这一原则贯彻落实了以人民为中心的理念。要坚持生态惠民、生态利民、生态为民，重点解决损害群众健康的突出环境问题，加快改善生态环境质量，提供更多优质生态产品，努力实现社会公平正义，不断满足人民日益增长的优美生态环境需要。

这一原则的实践需要依靠人民。坚持全民共治、源头防治，持续实施大气污染防治行动，打赢蓝天保卫战。加快水污染防治，实施流域环境和近岸海域综合治理。强化土壤污染管控和修复，加强农业面源污染防治，开展农村人居环境整治行动。加强固体废弃物和垃圾处置。要加强生态文明宣传教育，增强全民节约意识、环保意识、生态意识，培育生态道德和行为准则，开展全民绿色行动，营造爱护生态环境的良好风气，动员全社会都以实际行动减少能源资源消耗和污染排放，为生态环境保护作出贡献。

生态文明经典导读

节选

良好生态环境是最普惠的民生福祉。民之所好好之，民之所恶恶之。环境就是民生，青山就是美丽，蓝天也是幸福。发展经济是为了民生，保护生态环境同样也是为了民生。既要创造更多的物质财富和精神财富以满足人民日益增长的美好生活需要，也要提供更多优质生态产品以满足人民日益增长的优美生态环境需要。要坚持生态惠民、生态利民、生态为民，重点解决损害群众健康的突出环境问题，加快改善生态环境质量，提供更多优质生态产品，努力实现社会公平正义，不断满足人民日益增长的优美生态环境需要。

生态文明是人民群众共同参与共同建设共同享有的事业，要把建设美丽中国转化为全体人民自觉行动。每个人都是生态环境的保护者、建设者、受益者，没有哪个人是旁观者、局外人、批评家，谁也不能只说不做、置身事外。要增强全民节约意识、环保意识、生态意识，培育生态道德和行为准则，开展全民绿色行动，动员全社会都以实际行动减少能源资源消耗和污染排放，为生态环境保护做出贡献。

（节选自2018年5月习近平在全国环境保护大会上的讲话）

思考题

1. 阅读导读部分，划出最能体现"良好生态环境是最普惠的民生福祉"这一原则的特殊意义的语句并谈谈你对该句的理解。

2. 小组调查所在城市或家乡近三年雾霾、饮水安全、食品安全、垃圾围城等问题发生了什么变化，撰写调查报告或拍摄图片，并总结。

Section iv: Mountains, Rivers, Forests, Farmlands, Lakes and Grasslands Are a Community of Life

Introduction

1. Proposal process

On July 19, 2017, in the overall plan for the establishment of national park system deliberated and adopted at the 37th meeting of the central leading group for comprehensively deepening reform, the "grass" was incorporated into the same life community of mountains, rivers, forests, fields and lakes for the first time. On October 18, 2017, the report of the 19th National Congress of the Communist Party of China called for "overall management of landscape, forest, farmland, lake and grass system". So far, a systematic governance standard system including various environmental elements has been formed.

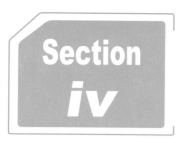

2. Spatial connotation

In the traditional spatial connotation, landscape belongs to one spatial system, forest, field and grass belong to another spatial system, and their indicators are counted separately, but "landscape, forest, field, lake and grass" is an organic natural ecosystem.

Mountains are divided into mountains and hills according to the height form. The altitude of mountains is more than 500 meters, which can be divided into extremely high mountains, high mountains, middle mountains and low mountains; by relative height, mountains can be divided into extremely undulating, large undulating, medium undulating and small undulating mountains. Hills are undulating terrain with a relative height difference of less than 200 meters. By relative height, they can be divided into gentle hills, low hills, medium hills and high hills.

Water includes rivers and lakes, which are divided into river corridors, lakes, reservoirs and wetlands according to the size of watershed area and water area. The river corridor can be divided into main stream, main tributary and other tributaries according to the drainage area. Forest includes woodland, shrub forest and other woodlands.

Fields generally refer to fields, including paddy fields, irrigated fields, dry lands, orchards, tea gardens and other gardens.

Grasslands includes natural grassland, artificial grassland and other grasslands. The space of forest, field and grass does not overlap with each other, but they depend on the mountains and rivers, and form an organic and orderly "life community" with mankind.

3. Significance and function

The lifeline of man lies in the field; the lifeline of the field lies in the water; the lifeline of water lies in the mountain; the lifeline of the mountain lies in the soil, and the lifeline of the soil lies in the forest and grass. This life community is the material basis for human survival and development. The life community between man and nature forms an interdependent and indispensable symbiotic and co-prosperity relationship. Only by taking "building a life community of mountains, rivers, forests, fields, lakes and grasses" as a major practical matter for people's livelihood and comprehensively protecting and repairing ecological resources such as mountains, rivers, forests, fields, lakes and grasses, can we continuously enhance the synergy and vitality of the life community and promote the construction of China's ecological civilization to a new level.

During the 13th Five Year Plan period, 25 pilot projects for ecological protection and restoration of mountains, rivers, forests, fields, lakes and grasses were carried out throughout the country. The 25 pilot projects involve 24 provinces and benefit 65 national poverty-stricken counties.

Excerpts

Mountains, rivers, forests, farmlands, lakes and grasslands are a community of life. An ecosystem is an integrated natural system of interdependent and closely-related ecological chains. The lifeline of humans rests with farmlands, that of farmlands with water, that of water with mountains, that of mountains with earth, and that of earth with forests and grasslands. This community of life is the material basis for the survival and development of humanity. We should take a broad and long-term view. We must try to avoid earning a little only to lose a lot, or attending to one thing and losing sight of others. Otherwise, our actions are bound to cause systematic and long-lasting damage.

(Excerpted from the speech delivered by Chinese President Xi Jinping at the National Conference on Eco-environmental Protection on May 18, 2018)

Vocabulary

natural system 生态系统 ecological chain 生态链

afforestation campaign 植树造林活动

Reflection and Activities

1. Introduce the typical cases of ecological restoration in China to the people around you by making English posters and Power Point.

Reference: xuexi qiang guo APP The typical cases of ecological restoration in China

2. Group tasks:

Read the project book of the first national college students' innovative design competition for ecological protection and restoration of "mountains, rivers, forests, lakes, grass and sand", and make an overview of the program description in English, including:

(1) Design name and definition

(2) Design background (analysis on improvement needs)

(3) Design scheme

(4) Design innovation

(5) Design feasibility demonstration

第四节 山水林田湖草是生命共同体

导读

1. 提出过程

2017年7月19日，中央全面深化改革领导小组第三十七次会议审议通过的《建立国家公园体制总体方案》中，首次将"草"纳入山水林田湖同一个生命共同体。2017年10月18日，党的十九大报告中要求"统筹山水林田湖草系统治理"。至此，形成了包括各环境要素在内的系统治理规范体系。

2. 空间内涵

在传统的空间内涵上，山水属于一个空间系统，林田草属于另一个空间系统，其指标均单独统计，但"山水林田湖草"是有机的自然生态系统。

山以高度形态为指标，分为山地和丘陵。山地的海拔高度在500米以上，其中又可分为极高山、高山、中山和低山；按相对高度来划分，可分为极大起伏的、大起伏的、中起伏的和小起伏的山地。丘陵则是相对高差在200米以下的起伏地形，按相对高度又可划分为缓丘陵、低丘陵、中丘陵以及高丘陵。

水包含河流及湖泊等，按流域面积和水域面积大小不同，分为河流廊道和湖泊水库湿地。其中河流廊道依据流域面积又可划分为干流、主要支流和其他支流。林包括有林地、灌木林地、其它林地。

田泛指田园，包括水田、水浇地、旱地、果园、茶园和其它园地。

草包括天然牧草地、人工牧草地和其它草地。林田草空间互不交叠，但是共同依存于山水之上，与人类共同组成了一个有机、有序的"生命共同体"。

3. 重大意义

人的命脉在田，田的命脉在水，水的命脉在山，山的命脉在土，土的命脉在林和草，这个生命共同体是人类生存发展的物质基础，人与自然生命共同体组成了相互依存、不可或缺的共生共荣关系。只有把"打造'山水林田湖草'生命共同体"作为重大民生实事紧紧抓在手上，对山水林田湖草等生态资源进行综合保护与修复，才能不断增强生命共同体的协同力和活力，推动我国生态文明建设迈上新台阶。

"十三五"期间，全国开展了25个山水林田湖草生态保护修复工程试点。这25个试点工程，涉及全国 24 个省份，惠及 65 个国家级贫困县。

节选

　　山、河、林、田、湖、草是生命共同体。生态系统是一个统一的自然系统，是相互依存、紧密联系的有机链条。人的命脉在田，田的命脉在水，水的命脉在山，山的命脉在土，土的命脉在林和草。这种生命共同体是人类生存和发展的物质基础。一定要算大账、算长远账、算整体账、算综合账，如果因小失大，顾此失彼，最终必然会对生态环境造成系统性、长期性破坏。

（节选自国家主席习近平 2018 年 5 月 18 日在全国生态环境保护会议上的讲话）

思考题

1. 制作英文海报、幻灯片，向你身边的人宣传中国生态修复典型案例。

参考资料：学习强国 APP《中国生态修复典型案例》

2. 小组任务：

阅读第一届全国大学生"山水林田湖草沙"生态保护与修复创新设计大赛参赛项目书，制作英文版项目书概述，包含：

（1）设计名称及定义

（2）设计背景（改进需求分析）

（3）设计方案

（4）设计创新点

（5）设计可行性论证

Section V: Protect the Ecological Environment with the Strictest System and the Strictest Rule of Law

Introduction

The fourth Plenary Session of the 19th CPC Central Committee included the system of environmental protection as an important part of upholding and improving the system of socialism with Chinese characteristics and modernizing China's governance system and capacity. It indicates that the institutional design of ecological civilization construction preliminarily completed since the 18th CPC National Congress is being internalized as an important part of the national governance system. To protect the ecological environment with the strictest system and the strictest rule of law is an important content of Xi Jinping's ecological Civilization thought. Only by implementing the strictest system and the strictest rule of law can we provide a reliable guarantee for the construction of ecological civilization.

The meaning of the strictest system

The strictest system is embodied in top-down dynamic management. It mainly includes 10 aspects: strict prevention, strict protection, strict standards, strictly prohibition, strict instructions, strict management, strict governance, strict investigation, strict prosecution, and severe punishment.

The meaning of the strictest rule of law

1. A strict legal system for environmental protection and pollution prevention and control should be established. The implementation of strict rule of law from the perspective of the normative object of ecological environment protection is mainly reflected in two aspects: one is the standardization of ecological environment protection and restoration; the other is the standard of environmental pollution prevention and control.

2. A strict system of legal norms corresponding to the behavior pattern of legal norms and legal consequences should be established.

3. A legal system covering all environmental elements of mountains, rivers, forests, fields, lakes and grass should be established.

4. A legal and normative system for governance at the source, systematically, comprehensively and in accordance with the law should be established.

5. An interlocking legal and normative system of legislation, law-abiding, law enforcement, and judicature should be established.

Excerpts

We should protect the eco-environment with the strictest regulations and laws. Eco-environmental protection relies on laws and regulations. Most of the outstanding environmental problems in China result from incomplete systems, the lack of appropriate rules and enforceable laws, inadequate implementation, and ineffective punishment. We must speed up institutional innovation, put in place more regulations, improve support systems, and strengthen their implementation. Regulations must be made mandatory, so that they become a powerful deterrent. We must use rules to supervise officials and their exercise of power in protecting blue skies and increasing green coverage. We must associate power with responsibility and accountability, so as to ensure the implementation of the CPC Central Committee's policies and plans for building an eco-civilization.

A country is strong when its law enforcement is strong; it is weak when its law enforcement is weak. The power of laws and decrees can only be established through their enforcement. The efficacy of rules lies in their implementation. We have introduced a series of reform measures and related rules. They must be carried out as strictly as the central inspection of environmental protection. The binding force and authority of the institutions must be firmly established, and selective implementation and perfunctory enforcement must be banned.

It is necessary to implement the accountability system among leading officials for protecting our eco-environment, and strictly assess their performance. Those who make ill-judged decisions that cause harmful consequences must be held accountable, and for life. We must never be lenient in punishing actions that damage our eco-environment. We will strike hard at typical cases that cause damage to the eco-environment and send out the signal that the perpetrators will be severely punished. Anyone who damages the environment—no matter where or when—shall face the consequences. The institutions must not become a "paper tiger".

(Excerpted from *Xi Jinping Governance of China III*)

Vocabulary

inspection 检查，视察

accountability system 问责制

mandatory 强制性的

law enforcement 执法

Reflection and Activities

In order to protect Yangtze River, China's first watershed protection law, "Yangtze River Protection Law of the People's Republic of China", was officially implemented on March 1, 2021.

1. Work in groups of four or five. Collect pictures and information of pollution in the Yangtze River. Discuss in groups why China has specially formulated a law to protect the Yangtze River.

2. Each of group members interviews three to five students on campus about their understanding of "We must strictly abide by the red line of ecological protection and implement the strictest system and the strictest rule of law".

3. Use the information your group has collected to prepare a report or PPT and make an oral presentation to the class.

第五节 用最严格制度、最严密法治保护生态环境

导读

党的十九届四中全会将生态环境保护制度列入坚持和完善中国特色社会主义制度、推进国家治理体系和治理能力现代化的重要内容。这标志着党的十八大以来初步完成的生态文明建设的制度设计正在内化为国家治理体系的重要组成部分。用最严格的制度、最严密的法治来保护生态环境是习近平生态文明思想的一个重要内容。只有实行最严格的制度、最严密的法治,才能为生态文明建设提供可靠保障。

"最严格的制度"

最严格的制度体现在自上而下的动态管理。主要包括十个方面:"严防""严保""严标""严禁""严谕""严管""严治""严查""严究""严惩"。

"最严密的法治"

一是建立严格的生态环境保护和污染防治法律规范体系。从生态环境保护的规范对象角度落实严密的法治主要体现在两个方面:一方面是对生态环境保护和恢复的规范。另一方面是对环境污染防治的规范。

二是建立法律规范的行为模式和法律后果相互对应的严密法律规范体系。

三是建立涵盖山水林田湖草各环境要素的法律规范体系。

四是建立源头治理、系统治理、综合治理、依法治理的法律规范体系。

五是建立立法、守法、执法、司法环环相扣的法律规范体系。

节选

用最严格制度最严密法治保护生态环境。保护生态环境必须依靠制度、依靠法治。我国生态环境保护中存在的突出问题大多同体制不健全、制度不严格、法治不严密、执行不到位、惩处不得力有关。要加快制度创新,增加制度供给,完善制度配套,强化制度执行,让制度成为刚性的约束和不可触碰的高压线。要严格用制度管权治吏、护蓝增绿,有权必有费、有贵必担当、失卖必追究,保证党中央关于性态文明建设决策部署落地生根见效。

奉法者强则国强，奉法者弱则国弱。令在必信，法在必行。制度的生命力在于执行，关键在真抓，靠的是严管。我们已出台一系列改革举措和相关制度，要像抓中央环境保护督察一样抓好落实。制度的刚性和权威必须牢固树立起来，不得作选择、搞变通、打折扣。

要落实领导干部生态文明建设责任制，严格考核问责。对那些不顾生态环境盲目决策、造成严重后果的人，必须追究其责任，而且应该终身追责。对破坏生态环境的行为不能手软，不能下不为例。要下大气力抓住破坏生态环境的反面典型，释放出严加惩处的强烈信号。对任何地方、任何时候、任何人，凡是需要追责的，必须一追到底，决不能让制度规定成为"没有牙齿的老虎"。

（节选自《习近平谈治国理政》第三卷）

思考题

为了保护长江，中国第一部流域保护法《中华人民共和国长江保护法》于2021年3月1日正式实施。

1. 四或五人一组，收集长江污染的图片和信息。小组讨论中国专门制定保护长江法律的意义。

2. 每小组成员采访3到5位同学，了解他们对"我们必须严格遵守生态保护的红线，实行最严格的制度和最严格的法治"的理解。

3. 用小组收集的信息准备报告或幻灯片，并向全班展示。

Section VI — Work Together to Promote a Global Eco-civilization

Introduction

Xi Jinping's thought of ecological civilization raised the ecological issue to the height of the development of human civilization, committed to forming a higher level of harmonious relationship between man and nature, based not only on the domestic, but also on the world, and based on the grasp of the law and general trend of social development. Putting forward the idea of building a global ecological civilization community, Xi Jinping's ecological civilization thought calls on all countries to participate and work together to jointly build a global ecological civilization, which provides Chinese wisdom and contributes Chinese strength to building a clean and beautiful world.

To build an ecological civilization, from an international perspective, we should take all mankind as the center, actively participate in and guide the reform of the global ecological environment governance system. We should not only build a beautiful China, but also do our best for the "beautiful world", set an example, contribute China's strength and provide China's solutions. The report of the 19th CPC National Congress clearly pointed out that we should become an important participant, contributor and leader in the construction of global ecological civilization. The COVID-19 epidemic sweeping the world has once again told people that the world is already an indivisible "life community" and has again made people rethink and examine the orientation of human beings in nature and the relationship between man and nature. Why does the world place high hopes and expectations on China? It is because China is a down-to-earth "doer". This action is not only reflected in the negotiation table, but also in the process of seeking common development and taking responsibility bravely. For example, China took the lead in issuing the National Plan on Implementation of the 2030 Agenda for Sustainable Development and the National Plan for Response to Climate Change (2014–2020). In addition, the Belt and Road Initiative advocated and promoted by our country is also an important "public benefit product" to promote the transformation of global ecological governance system and promote green development. The Belt and Road Initiative takes into account the interests of all parties, and is conducive to coordinating the rational utilization of resources in all countries in the world, promoting the development of green technologies and industries, and providing clean impetus for the sustainable development of the countries in the region. Only through joint construction can we share, and only through sharing can we communicate. The construction of global ecological civilization is an extension of the concept of a community with a shared future for mankind in the field of ecological construction, which is in line with the expectations of people all over the world for a clean and beautiful world. China's concept and practice on the construction of ecological civilization will surely leave a strong mark on the world's green and sustainable development.

China has become the world's largest country in energy conservation and utilization of new and renewable energy. China has been actively addressing climate change with a responsible attitude, taking climate change as a major opportunity to realize the transformation of development mode, and actively exploring a low-carbon development path in line with China's national conditions. The Chinese government has fully integrated addressing climate change into the overall strategy of national economic and social development. China has submitted its national independent contribution to the United Nations, which is not only aimed at promoting global climate governance, but also the internal requirement of China's development. It is the best effort that can be made to achieve the objectives of the Convention. China announced the establishment of a 20 billion yuan climate change South-South Cooperation Fund to support other developing countries. China is willing to continue to shoulder international responsibilities consistent with its national conditions, development stage and actual capabilities.

In order to build a global ecological civilization, China needs to deeply participate in global environmental governance. We should enhance China's voice and influence in the global environmental governance system, actively guide the reform direction of the international order, and form solutions for world environmental protection and sustainable development. We should take care of our home planet together, for ourselves and for future generations.

Excerpts

We should work together to promote a global eco-civilization. The eco-environment bears on the future of humanity. Building a green home is our common dream. Protecting the environment and dealing with climate change requires the joint efforts of all countries. No country can distance itself or remain immune from such challenges.

China has become an important participant, contributor and leader in promoting a global eco-civilization. We advocate jointly building a clean and beautiful world that respects nature and favors green development. China will be heavily involved in global environmental governance, have a bigger say and greater influence, play an active part in the transformation of the international order, and help form global solutions to eco-environmental protection and to sustainable development. We must always adopt the environment-friendly approach and play a constructive role in international cooperation on climate change. We will promote the philosophy and practice of eco-environmental progress in the Belt and Road Initiative to benefit the peoples of all countries along the Belt and Road.

(Excerpted from the speech delivered by Chinese President Xi Jinping at the National Conference on Eco-environmental Protection on May 18, 2018)

Vocabulary

global eco-civilization 全球生态文明   life community 生命共同体
renewable energy 可再生能源   the Belt and Road Initiative "一带一路"倡议

Reflection and Activities

1. Draw a mind map to list China's contributions to promoting global ecological civilization.

2. Referring to the passages in this section, translate the following paragraph into English.

在 G20 峰会（summit）之前，中国向联合国气候变化框架公约秘书处（secretariat of the UN Framework Convention on Climate Change）提交了两份关于其国家自主贡献（nationally determined contributions）和长期排放控制战略的文件。这是中国在执行《巴黎协定》方面的又一个具体行动，反映了中国追求绿色低碳发展的决心和努力。虽然中国决心与更广泛的世界一道减缓全球变暖，但西方国家应该承担起历史责任，履行国际义务。西方国家应对工业革命以来的大部分二氧化碳排放负有责任。他们还应该通过金融支持和技术转让来帮助发展中国家，而不是玩零和游戏（zero-sum game）。

第六节 共谋全球生态文明建设

导读

习近平的生态文明思想把生态问题上升到人类文明发展的高度，致力于形成更高层次的人与自然的和谐关系，不仅立足于国内，而且立足于世界，立足于把握社会发展的规律和大势，提出建设全球生态文明共同体的构想。习近平的生态文明思想要求各国共同参与、共同努力，共同建设全球生态文明，为中华民族提供智慧，为建设一个清洁美丽的世界贡献力量。

建设生态文明，从国际角度看，我们应该以全人类为中心，积极参与和指导全球生态环境治理体系的改革。我们不仅要建设一个美丽的中国，还要为"美丽的世界"尽最大努力，以身作则，贡献中国的力量，提供中国的解决方案。党的十九大报告明确指出，我们要成为全球生态文明建设的重要参与者、贡献者和领导者。新型冠状病毒流行病席卷全球，再次告诉人们，世界已经是一个不可分割的"生命共同体"，并再次引导人们反思和审视人类在自然和人与自然的关系中的定位。为什么世界对中国寄予厚望？这是因为中国是一个脚踏实地的"实干家"。这一行动不仅体现在谈判桌上，也体现在谋求共同发展、勇于承担责任的过程中。例如，中国率先发布了《中国落实2030年可持续发展议程国别方案》和《国家应对气候变化规划（2014—2020年）》。此外，我国倡导和推广的"一带一路"倡议，也是推动全球生态治理体系转型，促进绿色发展的重要"公益产品"。"一带一路"兼顾各方利益，有利于协调世界各国资源的合理利用，推动沿线绿色技术和产业发展，为沿线国家可持续发展提供清洁动力。只有共建才能共享，只有共享才能沟通。全球生态文明建设是人类共同未来共同体理念在生态建设领域的延伸，符合世界各国人民对清洁美好世界的期待。中国生态文明建设的理念和实践必将在世界绿色可持续发展中留下浓墨重彩的印记。

中国已成为世界上最大的节能和利用新能源和可再生能源的国家。中国以负责任的态度积极应对气候变化，以气候变化为重大机遇，实现发展方式转变，积极探索符合中国国情的低碳发展道路。中国政府已将应对气候变化全面纳入国民经济和社会发展总体战略。中国向联合国提交了国家独立捐款，这不仅是为了促进全球气候治理，也是中国发展的内在要求。这是为实现《公约》的目标所能作出的最大努力。中国宣布设立200亿元人民币的气候变化南南合作基金，以支持其他发展中国家。中国愿继续承担符合本国国情、发展阶段和实际能力的国际责任。

为了建设全球生态文明，中国需要深入参与全球环境治理。增强中国在全球环境治理体系中的话语权和影响力，积极引导国际秩序改革方向，形成世界环境保护和可持续发展的解决方案。我们应该共同保护地球家园，为我们自己和子孙后代。

节选

共谋全球生态文明建设。生态文明建设关乎人类未来，建设绿色家园是人类的共同梦想，保护生态环境、应对气候变化需要世界各国同舟共济、共同努力，任何一国都无法置身事外、独善其身。

我国已成为全球生态文明建设的重要参与者、贡献者、引领者，主张加快构筑尊崇自然、绿色发展的生态体系，共建清洁美丽的世界。要深度参与全球环境治理，增强我国在全球环境治理体系中的话语权和影响力，积极引导国际秩序变革方向，形成世界环境保护和可持续发展的解决方案。要坚持环境友好，引导应对气候变化国际合作。要推进"一带一路"建设，让生态文明的理念和实践造福沿线各国人民。

（节选自2018年5月18日中国国家主席习近平在全国生态环境保护大会上的讲话）

思考题

1. 根据以上新闻，制作一张思维导图，展示中国全球生态文明所做的贡献。
2. 参考本节所学内容，将以下这段话翻译成英文：

在G20峰会（summit）之前，中国向联合国气候变化框架公约秘书处（secretariat of the UN Framework Convention on Climate Change）提交了两份关于其国家自主贡献（nationally determined contributions）和长期排放控制战略的文件。这是中国在执行《巴黎协定》方面的又一个具体行动，反映了中国追求绿色低碳发展的决心和努力。虽然中国决心与更广泛的世界一道减缓全球变暖，但西方国家应该承担起历史责任，履行国际义务。西方国家应对工业革命以来的大部分二氧化碳排放负有责任。他们还应该通过金融支持和技术转让来帮助发展中国家，而不是玩零和游戏（zero-sum game）。

Chapter III

Amazing China

Chapter III selects eight typical cases of ecological restoration in China. It shows the typical practice of vigorously promoting ecological protection and restoration based on nature under the guidance of President Xi Jinping's thought on eco-civilization.

This chapter is divided into eight sections, namely Saihanba, Yucun, Mangrove to gold forest, Water, soil conservation—Changting, Fujian, Yunnan snub-nosed monkey range-wide conservation, Ecological restoration of Qingxi Country Park in Shanghai, Carbon peak and carbon neutrality and China embraces garbage classification, with Shanghai taking the lead. This chapter helps the students to accumulate knowledge, as well as improve their ability, and guides them to practice the idea of eco-civilization through practical activities, so as to achieve the goal of unity of knowledge and practice.

第三章

美丽中国

第三章选取中国生态修复典型案例中的八个案例,展现了在习近平生态文明思想的指导下,我国大力推动基于自然的生态保护修复工作的典型实践。

本章分为八小节,分别为塞罕坝、余村、红树林是金树林、福建长汀水土流失综合治理、滇金丝猴全境保护、上海青西郊野公园、碳中和碳达峰、中国采用垃圾分类——上海试点。本章逐步为学生搭建知识积累、能力提升的阶梯,最后引导学生通过实践活动,践行习近平生态文明思想,从而达到知行合一的目的。

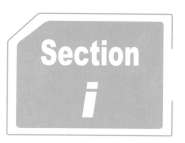

Saihanba

Introduction

Located in Chengde City, Hebei Province, northern China, the Saihanba National Forest Park stretches across 185,000 acres of forest. "Saihan" in the name is Mongolian, which means "beautiful"; "Ba" is Chinese, meaning "kaolin", and its full name can be translated as "beautiful kaolin".

Saihanba was once a royal retreat thanks to its cool summer weather and hunting area, however, the region turned into a desert by the end of the Qing Dynasty due to forest fires, deforestation and constant wars. Heavy northern winds from neighboring Inner Mongolia Autonomous Region worsened the situation.

The expansion of the desert also led to Beijing grappling with decades of sandstorms, which seriously threatened the capital's environment.

Over 350 foresters were first sent to the region to fight the desertification as early as the 1960s. Their duty was to rebuild a forest in Saihanba, but when they saw the extent of desertification in the region, nobody knew if the goal was realistic.

It was not until they found a 200-year-old larch, swaying alone in the wind, that their hope in their mission was rekindled.

After 55 years of efforts by three generations of experts, Saihanba has become the largest man-made forest park in the world.

In August 23, 2021, Xi Jinping came to Saihanba to observe the natural features of the forest farm, and to listen to Hebei's overall plan to promote the management of the grass and sand system and the management of the forest farm in the mountains and forests, learn about the promotion of Saihanba spirit in the forest farm.

Now, the forest absorbs 747,000 tons of carbon dioxide a year, which can produce a total transaction volume of more than 30 million yuan according to current prices on the Beijing carbon emissions trading market. The income will greatly help maintain the forest and improve north China's ecosystem.

Excerpts

Saihanba is considered to be a miracle as it has been transformed from barren land to lush forest through the extraordinary efforts of three generations of Chinese people.

Saihanba was previously been home to abundant forest resources and high biodiversity 400 years ago. With a cool summer and lush vegetation, the area was set to be a royal retreat. However, deforestation and constant wars turned the area into a desert by the end of the Qing Dynasty (1644–1911). As the forest barrier was gone, sandstorms became more frequent. To stop sandstorms that kept threatening or event striking Beijing, Tianjin and other northern China cities, the Forestry Administration decided in 1962 to set the Saihanba Mechanical Forest Farm, and sent 369 foresters, mostly in their 20s, to the area for tree planting.

The first group of foresters in Saihanba faced many challenges, equipped with only the simplest tools amid extreme coldness and drought. As result, they were unable to ensure the survival of trees planted.

However, after the joint efforts of three generations, Saihanba was restored and turned back into a green paradise with a forest coverage raised from 11.4 percent to 80 percent, which can conserve and purify 137 million cubic meters of water every year.

The miraculous planting story of foresters in Saihanba gives rise to the concept of the Saihanba Spirit, defined as working hard, advancing against difficulties, forging ahead, and innovating boldly. As one of the largest man-made plantations in the world, the Saihanba Afforestation Community won the Champions of the Earth award in 2017 due to the efforts to transform degraded land into a green paradise. Now, the lush Saihanba has become home to thousands of species of flora and fauna for its good environment, and also attracts numerous tourists.

(Excerpted from "Saihanba: The 'green miracle' created by three generations in N China")

Vocabulary

the Saihanba Spirit　塞罕坝精神

carbon emissions trading market　碳排放交易市场

Champions of the Earth award　地球卫士奖

Reflection and Activities

1. In 1962, 369 people with an average age of less than 24 from 18 provinces, autonomous regions and cities came to Chengde, Hebei Province to build Saihanba. In the past 60 years, three generations of Saihanba people have successfully turned the barren slope of "yellow sand covers the sky and the sun, birds have no habitat trees" into a forest sea of 10,000 mu with the courage and perseverance of "daring to teach the sun and the moon to change the sky". Today, the green "baton" has been passed to the "post-80s" and "post-90s" middle-aged and young forest farm workers. Closely following the footsteps of their predecessors, the "three generations of Lin" meet new opportunities and are not afraid of new challenges. While continuing the "green legend" of Saihanba, they play an exciting struggle movement with their youth.

Suppose you are a journalist who will come to Saihanba for an interview. What questions will you ask modern foresters in Saihanba?

(1) Documentary "Saihanba"

http://www.hebgcdy.com/ztbd/system/2017/10/09/030284295.shtml

(2) News report, "Young People of Saihanba"

2. There are currenthy about 900 national forest parks in China.

Investigate what national forest parks are there in your hometown. Write an English letter to introduce your foreign friends to the forest park in your hometown and invite them to visit.

第一节 1 塞罕坝

导读

塞罕坝国家森林公园位于中国北方河北省承德市，占地18.5万英亩。名称中的"塞罕"是蒙古语，意思是"美丽"；"坝"是中文，意思是"高岭"，其全名可以翻译为"美丽的高岭"。

塞罕坝因其凉爽的夏季天气和狩猎区一度是古代皇家的度假胜地，然而，由于森林火灾、森林砍伐和不断的战争，该地区在清末变成了沙漠。来自邻省内蒙古自治区的强烈北风也加剧了环境恶化。

沙漠的扩张还导致北京与几十年的沙尘暴作斗争，沙尘暴严重威胁着首都的环境。

早在20世纪60年代，就有350多名林业工作者被派往该地区抗击荒漠化。他们的职责是在塞罕坝重建一片森林，但当他们看到该地区荒漠化的程度时，没有人知道这个目标是否能现实。

直到他们发现一棵200年前的落叶松独自在风中摇曳，他们对使命的希望才重新燃起。

经过三代专家55年的努力，塞罕坝已成为世界上最大的人工森林公园。

2021年8月23日，习近平来到塞罕坝，观察林场的自然面貌，倾听河北的总体规划，促进沙草系统和山林区林场的管理，了解林场赛罕坝精神的推广。

目前，该森林每年吸收二氧化碳74.7万吨，按北京碳排放交易市场现行价格计算，总交易量可超过3000万元。这些收入将极大地帮助维护森林，改善中国北方的生态系统。

节选

塞罕坝被视为一个奇迹，因为在三代中国人的不懈努力下，它已经从不毛之地变成了郁郁葱葱的森林。

400年前，塞罕坝曾是丰富森林资源和高度生物多样性的家园。该地区夏季凉爽，植被茂盛，曾经是皇家避暑胜地。然而，到清朝末年（1644—1911），森林砍伐和连年战乱使该地区变成了沙漠。随着森林屏障的消失，沙尘暴变得更加频繁。为了阻止沙尘暴不断威胁、侵袭北京、天津和其他中国北方城市，国家林业局在1962年决定设立塞罕坝机械林场，并派出369名林务员（大多是20多岁的年轻人）到该地区植树造林。

塞罕坝的第一批林业工作者面临着许多挑战，在极端寒冷和干旱的环境中，他们只配备了最简单的工具。因此，他们无法确保种植的树木存活。然而，经过三代人的共同努力，塞罕坝得以恢复原貌，重新成为一个绿色天堂，森林覆盖率从11.4%提高到80%，每年可涵养和净化水源1.37亿立方米。

塞罕坝造林人的神奇造林故事，催生了塞罕坝精神，即艰苦奋斗、克服困难、开拓进取、大胆创新。作为世界上最大的人工造林地之一，塞罕坝造林社区凭借将退化土地转化为绿色天堂的努力，赢得了2017年"地球卫士"的荣誉。如今，郁郁葱葱的塞罕坝因其良好的环境而成为数千种动植物的家园，也吸引了众多游客。

（节选自中国国际电视台报道：塞罕坝：中国三代人创造的"绿色奇迹"）

思考题

1. 1962年，平均年龄不到24岁、来自全国18个省区市的369人来到河北省承德建设塞罕坝。近60年间，三代塞罕坝人以"敢教日月换新天"的气魄与毅力，成功将"黄沙遮天日，飞鸟无栖树"的荒坡变为万亩林海。如今，绿色"接力棒"已传递至"80后""90后"中青年林场职工手中。紧跟前辈的脚步，"林三代"们迎接新机遇、不畏新挑战，在续写塞罕坝"绿色传奇"的同时，用青春奏响激昂的奋斗乐章。

假如你是一名记者，来到如今的塞罕坝进行采访，你会问现代护林人们什么问题？

参考资料：

（1）纪录片《塞罕坝》

（2）新闻报道"塞罕坝的年轻人"

2. 中国目前有大约900座国家森林公园。调查你家乡所在的地区有哪些国家森林公园。写一封英文书信，向你的外国朋友介绍家乡的森林公园，并邀请他们来做客。

Section ii

Yucun

Introduction

Yucun Village in Anji County, Zhejiang Province is known for its lucid waters and lush mountains. At the entrance of a local park stands a towering monument bearing the inscription "blue waters and green mountains are indeed gold and silver mountains." This is where the "two mountains" theory was first put forward. As the cradle of the theory, Anji County is a fine example of ecological transformation.

Yucun village was honored as one of the "Best Tourism Villages" in the World Tourism Organization (UNWTO) General Assembly in Madrid on Dec 2, 2021.

In the 1990s, quarrying became a thriving business in Yucun. The mountains rich in high quality limestone were regarded as mountains of gold. In this county of less than 2,000 square kilometers, there were over 200 mining companies, or one in less than every eight square kilometers. The village became the largest limestone mining area in Anji, pocketing over three million yuan (around $458,000) in collective income on an annual basis.

However, the blooming mining business resulted in a severe damage to the local ecosystem. For a time, the village was clouded by smoke and fog, and plants were starting to become withered. Besides, residents were crippled and even dead because of mining accidents.

Villagers got rich by selling mineral products but found themselves in a poor ecological environment. Mountains were hollowed, waters tainted, and the air fouled.

The degrading environment and frequent mining accidents alarmed the villagers and they realized that an economic transformation was a must. But how should the polluting factories be closed down? How to protect the ecology and secure people's income? These questions brought the village into hesitation.

To address the ever deteriorating problem, Zhejiang province committed itself to building an "ecological province" in 2003.

Within three years, Yucun shut down three limestone quarries and a cement factory, which accounted for 95 percent of its annual income, causing much fear amongst the villagers. In traditional development models, what usually comes with economic growth are compromises in environmental protection.

In 2005, serving as secretary of the Zhejiang Provincial Committee of the CPC, Xi visited the village, where he proposed a guiding theory that endures to this day for the first time: "Lucid waters and lush mountains are invaluable assets."

The signature remark is part of Xi's thought on ecological civilization, which was formally established at the country's national conference on ecological protection in May 2018.

The thought is composed of several other key principles, including ensuring harmony between human and nature, regarding a sound ecological environment as the most inclusive benefits to people's wellbeing, seeing mountains, rivers, forests, farmlands, lakes, grasslands and deserts as a community of life, protecting the environment through the best institutional arrangements and the strictest rule of law, as well as getting deeply involved in global environmental governance to come up with a worldwide solution for environmental protection and sustainable development.

This has brought about tremendous changes in China.

The ensuing actions taken by the Chinese under the guidance of Xi's thought have strongly refuted those who thought they were hard-wired to see the trade-off between ecology and economy.

In traditional development models, what comes with economic growth are compromises in environmental protection.

If China kept its traditional and extensive mode of production, the country could hardly guarantee enough resources and environment for sustainable development.

Since the 18th CPC National Congress in 2012, China has revised its law on environmental protection, which was considered the strictest one in the country's history, and rolled out a series of documents to advance institutional reform on ecological progress.

From that time, environmental protection and ecological progress have been advanced in China regularly, institutionally, and systematically.

Guided by the "two mountains" theory, the villagers made the decision to conserve local environment and adjust their development planning. Leveraging its abundant bamboo resources, the village began to develop leisure eco-tourism including agritainment, homestay and rafting. The transformation has not only enabled high-quality economic development but also enhanced ecological conservation, industrial development and people's wellbeing. Yucun became a nationwide model in the endeavor to build a moderately prosperous society in all respects and beautiful countryside. This once seriously polluted village has become a hub of culture and tourism industries, known for the park commemorating the birth of the theory, a mine-remoulded park, countryside leisure resorts, a traditional Chinese medicine botanic garden, and a bamboo garden with jogging trails. Local industries have shifted from agriculture to mining and cement processing to bamboo products manufacturing, and to culture and tourism. More ways have been explored to turn green mountains into gold mountains, boost the village's economy and raise people's income.

Originating from this village, the "two mountains" theory has had a far-reaching and profound impact on China's ecological conservation. Sixteen years on, the phrase "clear waters and green mountains are indeed gold and silver mountains" has become an important part of Xi Jinping Thought on Socialism with Chinese Characteristics for a New Era. It is widely recognized among the Chinese people and has guided efforts for green development in every part of the country. Facing the new circumstances, tasks and requirements, it is the call of our times to follow the new philosophy of "accelerating reform of the system for developing an ecological civilization and building a beautiful China," and to advance ecological conservation, explore new paths to green development and earnestly implement the "two mountains" theory. This will open a new chapter of green development in the new era.

Recent years have witnessed Yucun Village's transformation from a mining town to a tourist destination. The village has witnessed the birth and development of the "two mountains" theory. Today, the village attracts numerous tourists and impresses friends from all over the world like Wang Mulin. It sets a fine example of implementing the "two mountains" theory and achieving economic development while protecting the environment.

Excerpts

At the beginning of the 1990s, Yucun, an idyllic mountain village in China's eastern Anji county, Zhejiang province, decided to end poverty by tapping the potential of its natural resources.

By mining for limestone and manufacturing cement, Yucun became one of China's richest villages with an annual revenue of more than 3 million yuan ($460,000).

However, it was not long before the villagers found it a Faustian bargain. Their hometown turned into a real eyesore with its pitted land, turbid rivers and dust haze.

Yucun was not the only village in Zhejiang that had seen its eco-system degrading. To address the ever deteriorating problem, the province committed itself to building an "ecological province" in 2003.

Within three years, Yucun shut down three limestone quarries and a cement factory, which accounted for 95 percent of its annual income, causing much fear amongst the villagers.

A heated debate emerged about the relationship between economic development and environmental protection—a relationship that was inevitably rocky during industrialization across the world.

In 2005, Xi, then secretary of the Zhejiang Provincial Committee of the Communist Party of China, visited Yucun. He assured the villagers that their move to close those factories was "wise."

"Mountains and rivers green are mountains of silver and gold," he said at the village's simple and humble meeting room.

Nine days later, in a commentary carried by Zhejiang Daily, Xi, using the pen-name "Zhe Xin," called for pursuing harmony between man and nature as well as harmony between the economy and society, in order to have clear water and green mountains along with "mountains of silver and gold."

The "two mountains" concept, later developed into Xi's thought on ecological civilization, has encouraged numerous Chinese cities and villages to pursue high-quality and sustainable growth through protecting the environment and developing green industries.

When Xi returned to Yucun 15 years later during an inspection tour in Zhejiang in March 2020, the village has transformed into a place featuring buildings with traditional white walls and black tiles, colorful flower fields and exquisite lotus ponds. The villagers made much more money than in the past thanks to a tourism boom.

"The path of green development is correct," Xi said.

"The path," wrote the United Nations Environment Programme in a report, "is beyond and does away with traditional development patterns and models, guiding the transformation of the production methods and the lifestyle of the entire society."

(Excerpted from "Green is gold—Xi Jinping innovates fight against climate change" on *China Daily*)

Vocabulary

Lucid waters and lush mountains are invaluable assets.　绿水青山就是金山银山。
the 18th CPC National Congress　中国共产党第十八次全国代表大会
"two mountains" theory　两山论

Reflection and Activities

Supposing you are a journalist and are interviewing a villager of Yucun, report the recent changes and make a vlog.

第二节 余村

导读

浙江省安吉县余村以清澈的水和郁郁葱葱的山而闻名。在当地一个公园的入口处矗立着一座高耸的纪念碑，上面刻着"绿水青山真是金山银山"，这就是"两座山"理论最早提出的地方。作为这一理论的摇篮，安吉县是生态转型的典范。

2021年12月2日，在马德里举行的世界旅游组织（UNWTO）大会上，余村被评为"最佳旅游村"之一。

20世纪90年代，采石业成为该地区一项蓬勃发展的行业。富含优质石灰岩的山脉被认为是金山。在这个面积不到2000平方公里的县，有200多家矿业公司，即每不到8平方公里就有一家。该村成为安吉最大的石灰岩矿区，每年的总收入超过300万元（约45.8万美元）。

然而，蓬勃发展的采矿业对当地生态系统造成了严重破坏。村庄一度被烟雾笼罩，植物开始枯萎。此外，居民因采矿事故而致残，甚至死亡。

村民们靠卖矿产品致富了，但生态环境极为恶劣。山被挖空了，水腐臭了，空气被污染了。

不断恶化的环境和频繁发生的矿难让村民们警觉起来，他们意识到必须进行经济转型。但是污染工厂应该如何关闭呢？如何保护生态，保障人民收入？这些问题使全村人犹豫不决。

为了解决日益恶化的问题，浙江省于2003年承诺建设"生态省"。

在三年内，余村关闭了三座石灰石采石场和一座水泥厂，这些采石场和水泥厂占其年收入的95%，在村民中引起了极大的恐慌。在传统的发展模式中，经济增长带来的是环境保护方面的妥协。

2005年，习近平担任浙江省委书记期间在安吉调研，在那里他首次提出了一个指导理论，坚持至今："绿水青山就是金山银山。"

这一标志性言论是习近平生态文明思想的一部分，该思想在2018年5月的全国生态保护大会上正式确立。

该思想还包括其他几个关键原则，包括确保人与自然的和谐，将良好的生态环境视为对人民福祉最具包容性的利益，将山、河、林、田、湖、草原和沙漠视为一个生命共同体，通过最佳制度安排和最严格的法治来保护环境，并深入参与全球环境治理，为环境保护和可持续发展提出全球解决方案。

这给中国带来了巨大的变化。

在习近平思想的指导下，中国采取的行动有力地驳斥了那些认为中国人在生态和经济之间难以取舍的观点。

在传统的发展模式中，经济增长带来的是环境保护方面的妥协。

如果中国保持传统的粗放型生产方式，就很难保证有足够的资源和环境来实现可持续发展。

自2012年党的十八大以来，中国修订并颁布了中国历史上最严格的环境保护法，并出台了一系列文件推进生态建设体制改革。

从那时起，环境保护和生态进步在中国得到了定期的、制度化和系统化的推进。

在"两山论"的指导下，村民们做出了保护当地环境和调整发展规划的决定。他们利用丰富的竹子资源，该村开始发展休闲生态旅游，包括农家乐、民宿和漂流。这一转型不仅实现了高质量的经济发展，还加强了生态保护、工业发展和人民福祉。余村成为全国全面建设小康社会和美丽农村的模范。这座曾经污染严重的村庄已成为文化和旅游业的中心，以纪念"理论"诞生的公园、矿山改造公园、乡村休闲度假村、中药植物园和带慢跑道的竹子园而闻名。当地工业已从农业转向采矿和水泥加工，再转向竹制品制造，再转向文化和旅游业。探索了更多的途径，将青山变成金山，促进村庄经济发展，提高人民收入。

"两山论"源于这个村庄，对中国的生态保护产生了深远的影响。十六年来，"绿水青山就是金山银山"已成为习近平新时代中国特色社会主义思想的重要组成部分。它得到了中国人民的广泛认可，并指导了全国各地的绿色发展努力。面对新的形势、任务和要求，遵循"加快生态文明建设体制改革，建设美丽中国"的新理念，推进生态保护，探索绿色发展新途径，认真贯彻"双山"理论，是时代的呼唤。这将开启新时代绿色发展的新篇章。

节选

20世纪90年代初，位于中国浙江省安吉县东部的一个田园山村余村决定通过挖掘其自然资源的潜力来结束贫困。

通过开采石灰石和生产水泥，余村成为中国最富有的村庄之一，年收入超过300万元（46万美元）。

然而，村民们不久就发现这是一笔浮士德式的交易。他们的家乡变得坑坑洼洼、河水混浊、尘土飞扬。

余村并不是浙江唯一一个生态系统退化的村庄。为了解决日益恶化的问题，该省于2003年开始致力于建设"生态省"。

在三年内，余村关闭了三座采石场和一座水泥厂，这些采石场和水泥厂占了余村年收入的95%，在村民中引起了极大的恐惧。

关于经济发展和环境保护之间关系的激烈辩论随之展开——在全球工业化进程中，这一关系无疑会遭遇困难。

2005年，时任中共浙江省委书记的习近平视察了余村。他向村民保证，他们关闭这些工厂的举动是"明智的"。

他在村里简陋的会议室里说："青山绿水是金山和银山。"

九天之后，习近平以"哲欣"为笔名，在《浙江日报》发表了一篇评论，呼吁追求人与自然的和谐，经济与社会的和谐，既要"金山银山"也要"绿水青山"。

"两山"理念后来发展成为习近平生态文明思想，鼓励中国众多城市和乡村通过保护环境和发展绿色产业追求高质量和可持续的增长。

15年后的2020年3月，当习近平在浙江考察时再次来到余村，这个村子变了模样，白墙、黑瓦的传统建筑林立、五彩缤纷的花圃和玲珑的莲花池遍布。得益于旅游业的繁荣，村民们赚的钱比过去多得多。

"绿色发展的道路是正确的，"主席说。

联合国环境规划署在一份报告中写道："这条道路超越并摒弃了传统的发展模式，引导着整个社会生产方式和生活方式的转变。"

（节选自《中国日报》报道：绿水青山就是金山银山：习近平创新应对气候变化）

思考题

假设你是一名记者，对余村的村民进行一次采访，对余村近年的变化做一次报道，并制作视频网络日志。

Section iii

Mangrove to Gold Forest

Introduction

Zhanjiang Mangrove Reserve was founded as a provincial reserve in 1990 and upgraded to a national nature reserve with the approval of the State Council in 1997. The total protected area is 20,278.8 hectares, including more than 9000 hectares of natural mangroves, accounting for 33% of the total mangrove area in China and 79% of the total mangrove area in Guangdong Province. It is the largest coastal mangrove nature reserve in mainland China. It is a forest and wetland type nature reserve. The main protection objects are tropical mangrove wetland ecosystem and its biodiversity, including mangrove resources, adjacent beaches, water surfaces and wild animals living in the forest. In January 2002, the reserve was listed in the list of international important wetlands under the "Ramsar Convention", becoming a key area for biodiversity protection in China and an important base for in-situ protection of international wetland ecosystem. In 2005, it was identified as a national wild animal (bird) epidemic focus and disease monitoring point and a national coastal shelter forest monitoring point.

Zhanjiang Mangrove Reserve is rich in natural resources. There are 24 species of true mangrove and semi mangrove plants in 15 families, and 21 species of main associated plants in 14 families. It is the area with the most mangrove species on the coast of mainland China. Among them, the most widely distributed and the largest number are *Aegiceras corniculatum*, *Avicennia marina*, *Rhizophora stylosa*, *Kandelin canda* and mangrove. The main forest vegetation communities include *Aegiceras corniculatum*, *Avicennia marina*, *Kandelin canda*, *Rhizophora stylosa* pure forest community and *Aegiceras corniculatum*, *Avicennia marina*, *Avicennia marina*, *Kandelin canda*, *Avicennia marina* + *Rhizophora stylosa* and other communities, and the canopy closure of the forest is more than 0.8. There are 194 species of birds recorded, making this place an important bird area in Guangdong. In addition, there are 130 species of shellfish in 76 genera, 41 families, 3 classes, and 139 species of fish in 100 genera, 60 families, 15 orders.

Excerpts

Mangrove ecological protection and restoration have faced many problems. For some time, many areas where mangroves originally grew have been transformed into fish ponds, farmland or land for infrastructure construction. At the same time, affected by the invasion of alien species such as Spartina, alterniflora and Coriaria, apetala, and the direct discharge of production and domestic sewage, the mangrove area in Zhanjiang, Guangdong Province has been continuously reduced, the habitat has been fragmented, and the ecosystem has been damaged and degraded.

In recent years, the state has issued a number of regulations and measures related to mangrove protection and restoration, and Guangdong Zhanjiang Mangrove National Nature Reserve Administration has also continuously strengthened its management, and mangrove protection and restoration has achieved positive results. However, due to the large area of management area, scattered management units and insufficient management personnel, the reserve still faces many problems, especially the pressure of returning ponds to forests, weak management and protection force in the later stage, lack of funds and other problems, which affect the final effect of mangrove protection and restoration.

Mangrove forests account for only a small percentage of the total forest area in China, but they have attracted the attention of Chinese leaders. On April 19, 2017, President Xi Jinping urged protection of mangrove forests while inspecting the Golden Gulf Mangrove Ecological Protection Zone in Beihai city in the Guangxi Zhuang autonomous region. And, on 8 June 2020, the theme of World Oceans Day in China was set as "Protecting mangroves, Protecting Ocean Ecosystems", indicating the government's efforts to increase public recognition of the valuable functions of mangroves. A social climate conducive to the conservation and restoration of mangroves is emerging in China.

On June 8, 2021, Guangdong Zhanjiang Mangrove National Nature Reserve Administration, the third Marine Research Institute of the Ministry of natural resources and Beijing entrepreneur Environmental Protection Foundation jointly signed the first 5,880 ton carbon emission reduction transfer agreement of "Zhanjiang Mangrove afforestation project", marking the formal completion of China's first "blue carbon" trading project. This provides a demonstration for the way to realize the value of ecological products of mangrove and other blue carbon ecosystems, and is of great significance in encouraging social capital investment in mangrove ecological protection and restoration and boosting the realization of carbon neutralization.

(Excerpted from "Rare bird species land to roost in Zhanjiang" from *China Daily*)

Vocabulary

national geopark 国家地质公园　　　　mangrove 红树林
wetland 湿地　　　　　　　　　　　　ecological red line zones 生态红线区域
nature reserve 自然保护区

Reflection and Activities

1. Please list a National Nature Reserve in the province where your hometown is located and make an introduction to PPT or video to introduce the beautiful environment of your hometown.

2. Which area of your hometown do you think needs to be improved? Please design an improvement plan.

第三节 3 红树林是金树林

导读

湛江红树林保护区是1990年建成的省级自然保护区，在1997年在国务院批准下升级为国家级自然保护区。保护区总面积达20278.8亩，其中包括9000亩天然红树林，占全国红树林总面积的33%，占广东省红树林总面积的79%。这是中国大陆面积最大的沿海红树林自然保护区。它是一个森林和湿地交错的自然保护区。保护区主要的保护目标是热带红树林湿地生态系统及其多样性，包括红树林资源、临近滩涂、水面及栖息于林内的野生动物。2002年1月，该保护区被"拉萨姆公约"列入全球重点湿地保护项目，成为我国生物多样性保护的关键性地区和国际湿地生态系统就地保护的重要基地。2005年被确定为国家级野生动物（鸟类）疫源疫病监测点、国家级沿海防护林监测点。

湛江红树林保护区自然资源十分丰富。有真红树和半红树植物15科24种，主要的伴生植物14科21种，是我国大陆海岸红树林种类最多的地区。其中分布最广、数量最多的为白骨壤、桐花树、红海榄、秋茄和木榄，主要森林植被群落有白骨壤、桐花树、秋茄、红海榄纯林群落和白骨壤+桐花树、桐花树+秋茄、桐花树+红海榄等群落，林分郁闭度在0.8以上。记录有鸟类达194种，是广东省重要鸟区之一。此外，贝类有3纲41科76属130种，鱼类有15目60科100属139种。

节选

红树林生态保护与修复曾面临诸多难题。一段时间以来，当地许多原来生长有红树林的区域被改造为养殖鱼塘、农田或者基础设施建设用地等其他用途，同时，受互花米草、无瓣海桑等外来物种入侵和生产生活污水直接排入等影响，广东湛江红树林面积持续减小、生境破碎化、生态系统受损退化。

近年来，国家出台多项红树林保护修复相关的规定措施，广东湛江红树林国家级自然保护区管理局也不断加大管理力度，红树林保护修复取得积极成效。但是由于管理区域面积大、管理单元分散、管理人手不足等问题，保护区仍然面临许多难题，尤其是退塘还林压力、后期管护力量薄弱、资金缺乏等问题，这些都影响红树林保护修复的最终效果。

红树林只占中国森林总面积的一小部分，但它们已经引起了中国领导人的高度重视。2017年

4月，习近平主席在考察广西壮族自治区北海市金湾红树林生态保护区的同时，呼吁保护红树林。2020年6月8日，中国世界海洋日的主题定为"保护红树林，保护海洋生态系统"，表明政府努力提高公众对红树林宝贵功能的认识。中国正在形成有利于保护和恢复红树林的社会气候。

2021年6月8日，广东湛江红树林国家级自然保护区管理局、自然资源部第三海洋研究所和北京市企业家环保基金会，共同签署"湛江红树林造林项目"首笔5880吨的碳减排量转让协议，标志着我国首个"蓝碳"交易项目正式完成。这为红树林等蓝碳生态系统的生态产品价值实现途径提供了示范，在鼓励社会资本投入红树林生态保护修复、助推实现碳中和方面具有重要意义。

（节选自《中国日报》报道：珍稀鸟类在湛江栖息）

思考题

1. 同学们请列举一个自己家乡所在省份的国家级自然保护区，做一个介绍PPT或视频来宣传家乡的美丽环境。

2. 你觉得你家乡的哪个区域环境亟待改进？请设计一个改进方案。

Water, Soil Conservation—Changting, Fujian

Introduction

General introduction

Changting, known as "Tingzhou" in ancient times, is located in the west of Fujian and at the southern foot of the Wuyi Mountains. The county governs 18 townships and 300 villages, with a total population of 550,000 and a land area of 3,104.16 square kilometers. It is the fifth largest county in Fujian Province.

Changting is one of the country's old revolutionary base areas. These local strongholds were established in remote areas with mountainous or forested terrain by revolutionary forces led by the Communist Party of China as it fought the Kuomintang during the Land Revolution Period (1927-1937) and the War of Resistance against Japanese Aggression (1931-1945).

Past and Present

The mountainous village used to experience extremely serious soil erosion and water loss, as did the entire county. A survey conducted in 1985 by the local government found that nearly 100,000 hectares of land in Changting were subject to soil erosion—one-third of the county's total area. The exposed red earth earned the area the nickname "flaming mountain". The county was featured with bare mountains, muddy water, barren land and poverty-stricken people.

Remote sensing data in 1985 showed that the county had about 98,000 hectares subject to soil erosion, accounting for 31.5 percent of its land area. When soil erosion is at its worst, rainfall can cut hillsides into fragments that will develop into larger gullies over time. Along the broken and barren mountains, rainwater will carry sediment into farmland and block waterways at the foot of the mountains, causing floods and damaging crops.

"The mountains of Changting are all red, shining like blood... There are no insects, no rat trails, no roosting birds. Nothing but a miserable silence, always accompanied by the destroyed spirit of the mountains."

By the end of 2020, only about 21,000 hectares in Changting still had soil erosion problems, a decline of about 78 percent, according to the local government. The proportion of forest coverage in Changting reached 80 percent, with 16.23 million square kilometers of forest and 275.35 square kilometers of nature reserve, according to the local authorities.

Changting has become a success story by persevering with green development as China advances the building an ecological civilization as part of the national development strategy. So much so that the measures employed were included in a compendium of typical cases for the 15th meeting of the Conference of the Parties to the Convention on Biological Diversity, also known as COP 15, held in Kunming, Yunnan Province, in October. Changting's residents have found a green path to riches by expending large amounts of time, energy and money to overcome their problems.

The embodiment of Xi Jinping's thoughts

Xi Jinping profoundly pointed out: "ecological resources are the most precious resources in Fujian, ecological advantages are the most competitive advantages of Fujian, and the construction of ecological civilization should be the construction that Fujian should pay the most attention to." The construction of ecological civilization in Fujian started early and made great efforts. As early as 2000, Xi complied with the expectations of the people, put forward the strategy of building an "ecological province" with great foresight, personally served as the leader of the leading group for the construction of an ecological province, and launched the most systematic and largest environmental protection action in Fujian.

After nearly 30 years of unremitting efforts, Changting County in Fujian Province, which used to be mountainous, muddy, thin and poor, has become a brand of soil erosion control in China and a banner of soil erosion control in the south, The unity of ecological beauty and people's wealth has been realized, fully embodying comrade Xi's correct thought that "clear water and green mountains are golden mountains and silver mountains". The governance mode with regional characteristics formed by Changting in soil

and water conservation and ecological construction has been highly praised by comrade Xi and is called "Changting experience". Comrade Xi Jinping attaches great importance to environmental protection and ecological governance. As early as more than a decade ago, he began to pay attention to the problem of soil erosion in Changting, and within less than a month at the end of 2011 and the beginning of 2012, he made important instructions on soil erosion control in Changting twice. Comrade Xi Jinping's two "Changting instructions" were put forward under the background of China's reform and opening-up and ecological civilization construction entering a new stage, and new achievements in the construction of Fujian Ecological Province. They have unique scientific connotations.

The great achievements made in the comprehensive control of soil and water loss and the construction of ecological civilization in Changting are the vivid practice of Xi's thought on ecological civilization, which provides a new model from ecological restoration, ecological poverty alleviation to ecological revitalization for areas suffering from soil and water loss. One of them is to stimulate the core power of green transformation based on concept innovation. From logging for heating, burning mountains and destroying forests, to viewing green water and green mountains as invaluable assets, and then to transforming them into invaluable assets, Changting has experienced great changes in ideology, gradually established correct ecological values and realized a complete change in the mode of development.

Reward and influences

People in Changting firmly establish the concept of "lucid waters and lush mountains are invaluable assets", and continue to carry out soil erosion control in the spirit of "dripping water wears through the rock stone".

Remarkable results have been achieved in ecological governance. In 2020, the total area of soil erosion in Changting County decreased by 760.84 square kilometers, the soil erosion rate reduced to 6.78%, the forest coverage rate increased to 80.3%, the forest volume increased to 17.79 million cubic meters, and the wetland area reached 3,513 hectares. The air quality in Changting remains above the standard grade II all year round, the water quality rate of national and provincial standards and drinking water reached 100%.

Biodiversity has been restored. Vascular plants increased from 110 species before treatment to 340 species now; bird species increased from less than 100 species to 306 species; rare and endangered wild animals such as white necked long tailed pheasant, yellow bellied pheasant, Sumen antelope and leopard cat have also returned to the mountains and forests.

Green industry promotes development. In 2020, Changting County received a total of 1 million tourists, achieved an annual output value of 1.166 billion yuan, and the per capita disposable income of farmers reached 18,149 yuan, with an average annual growth of 10.7%. From 2017 to 2020, it was awarded the title of "top ten counties in County economic development in Fujian Province" for four consecutive years.

Excerpts

Changting, a county in Fujian Province, is said to possess "red genes" from the time it served as a Red Army base in the early 1930s. However, over the past two decades the county has also emerged as a role model in pursuing green development after resolving a decades-old soil erosion problem.

Changting has a well-earned reputation for its efforts to control water loss and soil erosion. President Xi Jinping has called for efforts to tackle such problems to be advanced nationwide by using the "Chanting experience", as China strives to build an ecological civilization in which sustainable development is key.

In 1983, efforts to curb environmental damage in Changting were stepped up as the Fujian provincial government started to control and curb soil erosion and the county became a trial area for such efforts.

When Xi Jinping worked in Fujian in various positions from 1987 to 2002, he placed great emphasis on conservation.

He said that in Fujian, "lush mountains and lucid waters are priceless assets". This has become an important part of his governance philosophy for building an ecological civilization with a view to creating a fruitful balance between economic development and environmental protection, and harmony between humans and nature.

Xi traveled to Changting on five occasions, leading the local government in tackling soil and water damage. He told officials efforts had been made to control such damage, even in the days of old China. "Now that the country is governed by the CPC, we are able to do it better," he said.

In November, 1999, Xi, who was deputy secretary of the CPC Fujian Provincial Committee and acting governor of the province, launched a campaign to harness water loss and soil erosion. In February the following year, the campaign was listed among 15 projects benefiting people in the province the most.

Vocabulary

gene 基因
curb 控制
deputy 副手；副职

soil erosion 水土流失
conservation 保护

Reflection and Activities

1. Look up into relevant reports and find out about Xi jinping's efforts to improve Changting's environment during his stay in Fujian Province, and talk about your understanding of the thought of ecological civilization in the new era.

2. Make illustration and a brief report on the development history of ecological governance in Changting, highlighting the comparative changes of people's life and environment before and after the improvement of soil and water loss.

第四节 中国水土流失综合治理——福建长汀

导读

概述

长汀，古称"汀州"，位于福建西部，武夷山南麓。全县辖18个乡（镇）、300个村（居），总人口55万，面积3104.16平方公里。是福建省第五大县。

长汀是全国革命老区之一。这些地方据点是由中国共产党领导的革命力量在土地革命时期（1927—1937）和抗日战争（1931—1945）期间与国民党作战时在山区或森林地形的偏远地区所建。

过去和现状

这个山村曾经水土流失极为严重，整个县城也是如此。当地政府在1985年进行的一项调查发现，长汀有近10万公顷土地遭受水土流失，占全县总面积的三分之一。裸露的红土为这里冠上了"火焰山"的绰号。该县山秃、水浑、土地贫瘠、人民困苦。

1985年遥感资料显示，全县水土流失面积约9.8万公顷，占国土面积的31.5%。当土壤侵蚀最严重时，降雨会将山坡划得支离破碎，随着时间的推移这些缝隙会发展成更大的沟壑，沿着破碎荒山，雨水会将泥沙带入农田，堵塞山脚下的水道，引发洪水，毁坏庄稼。

"长汀的群山都是红色的，像血一样闪耀……没有虫子，没有老鼠的踪迹，没有栖息的鸟儿。只有凄惨的寂静，总是伴随着被毁坏的山灵。"

据当地政府统计，截至2020年底，长汀仅约2.1万公顷土地仍存在水土流失问题，下降约78%。长汀市森林覆盖率达到80%，森林面积1623万平方公里，自然保护区面积275.35平方公里。

长汀坚持绿色发展，已成为中国将生态文明建设纳入国家发展战略的成功典范。10月，在云南省昆明市举行的《生物多样性公约》缔约方大会第十五次会议（即第十五届缔约方会议）的典型案例汇编中包含了长汀所采取的措施。长汀人民花了大量的时间、精力和金钱来解决他们的问题，找到了一条绿色的致富之路。

习近平生态文明思想

总书记深刻指出："生态资源是福建最宝贵的资源，生态优势是福建最具竞争力的优势，生态文明建设应当是福建最花力气抓的建设。"福建生态文明建设起步早、力度大，早在2000年，总书记就顺应人民群众期盼，极具前瞻性地提出了建设"生态省"战略，亲自担任生态省建设领导小组组长，开启了福建最为系统、最大规模的环境保护行动。

曾经山光、水浊、田瘦、人穷的福建省长汀县，经过近三十年的不懈努力，一跃成为中国水土流失治理的品牌、南方水土流失治理的一面旗帜，实现了生态美与百姓富的统一，充分体现了习近平同志所说的"绿水青山就是金山银山"这一思想的正确性。长汀在水土保持与生态建设方面形成的具有地域特色的治理模式得到了习近平同志的高度赞扬，并称之为"长汀经验"。习近平同志对环境保护与生态治理工作极为重视，早在十多年前，他就开始关注长汀的水土流失问题，并于2011年底和2012年初不到一个月的时间内，先后两次对长汀的水土流失治理工作作出了重要批示。习近平同志的两次"长汀批示"是在中国改革开放与生态文明建设进入新阶段、福建生态省建设取得新成就的背景下提出的，具有独特的科学内涵。

长汀水土流失综合治理和生态文明建设取得的巨大成就是习近平生态文明思想的生动实践，为饱受水土流失之苦的地区提供了从生态恢复、生态脱贫到生态振兴的新模式。其中之一就是以观念创新为本，激发实现绿色转型的内核动力。从伐木取暖、烧山毁林到视绿水青山为金山银山，再到将绿水青山转化为金山银山，长汀经历了思想上的巨变，逐步确立了正确的生态价值观，实现了发展方式的彻底变革。

回报与影响

长汀人民牢固树立"绿水青山就是金山银山"的理念，以"滴水穿石、人一我十"的精神，持续开展水土流失治理。

生态治理取得明显成效。2020年，福建省长汀县累计减少水土流失面积760.84平方公里，水土流失率降为6.78%，森林覆盖率提高到80.3%，森林蓄积量提高到1779万立方米，湿地面积达3513公顷，空气环境质量常年维持在Ⅱ级标准以上，国、省控断面水质和饮用水源地水质达标率均为100%。

生物多样性得到恢复。维管植物数量从治理前的110种增加到340种；鸟类从不到100种恢复到306种，白颈长尾雉、黄腹角雉、苏门羚、豹猫等珍稀濒危野生动物也纷纷重新回到山林。

绿色产业促发展。2020年，福建省长汀县共接待游客100万人次，实现年产值11.66亿元，农民人均可支配收入达18149元，年均增长10.7%，2017年至2020年连续四年荣膺"福建省县域经济发展十佳县"称号。

节选

福建省长汀县在20世纪三十年代初曾是红军根据地，因此被认为拥有"红色基因"。然而，在过去的二十年里，该县在解决了数十年的水土流失问题后，也成为了追求绿色发展的典范。

长汀在控制水土流失方面的努力享有盛誉。

在中国建设以可持续发展为核心，建设生态文明之际，习近平主席呼吁以"唱诵经验"在全国范围内推进解决此类问题。

1983 年，福建省政府启动水土流失防治工作，长汀县成为环境污染防治试点地区。

1987 年至 2002 年，习近平在福建担任不同职务时，都十分重视保护工作。

他说，在福建，"青山绿水是无价之宝"。这已成为他建设生态文明治理理念的重要组成部分，旨在实现经济发展与环境保护、人与自然和谐相处的卓有成效的平衡。

思考题

1. 查阅有关报道，了解习近平在福建期间为改善长汀环境所做的努力，谈谈你对新时代生态文明思想的理解。

2. 对长汀生态治理的发展历程进行图解和简报，突出水土流失改善前后人们生活环境变化比较。

Section V

Yunnan Snub-nosed Monkey Range-wide Conservation

A female Yunnan snub-nosed monkey and her two infants are seen in the Baima Snow Mountain Nature Reserve in Yunnan province. Wang Changshan/Xinhua

Introduction

Yunnan snub-nosed monkey is a unique endangered species in China. It is can be only found in the 7,000 square kilometer high-altitude forest between the Lancang River and the Jinsha River at the junction of Tibet and Yunnan. The three parallel rivers world natural heritage site where the habitat belongs has 20% of China's higher plants and 25% of vertebrates. It is one of the globally recognized biodiversity models.

The global protection of Yunnan snub-nosed monkeys and its habitat helps to maintain the integrity of regional primitive forest ecosystem and rich biodiversity, and plays an important role in building an ecological security barrier in southwestern China.

1. Challenges

In the past 40 years, the protection of Yunnan snub-nosed monkeys has made great progress with the joint efforts of the government and social forces. The number has increased from more than 1,500 in the first survey in the 1990s to more than 3,000 at present.

However, considering the landscape scale of the whole distribution area, it still faces many challenges. Insufficient understanding of biodiversity conservation and management, such as lack of unified monitoring of biodiversity still exists; some key habitats have not been incorporated into the nature reserve system and lack effective management and protection; roads, villages and farmland cause habitat fragmentation and block gene exchange among monkeys; local people are highly dependent on production methods such as planting, gathering and grazing under the forest, and there is a conflict between local economic development and ecological protection; the protection investment does not match the demand, and the public awareness and participation are low.

2. Joint protection mechanisms

Since 2019, 13 institutions including Yunnan Forestry and Grassland Bureau, Yunnan Green Environment Development Foundation and The Nature Conservancy have jointly launched the "Yunnan Snub-nosed Monkey Range-wide Conservation network", trying to establish a multi-party joint protection mechanism and take actions from the aspects of patrol monitoring, corridor restoration, community participation, friendly development and public participation, so as to achieve the sustained and healthy growth of Yunnan snub-nosed monkey population. The health level of habitat ecosystem has been comprehensively improved.

Standardization and informatization of biodiversity patrol monitoring.

Several standardized monitoring plans and the online data management platform and mobile app for patrol monitoring have come into full play in the patrol monitoring, thus to realize the paperless, information-based and automatic collection, storage and analysis of patrol monitoring data.

Establish community protection areas to cover protection vacancies.

The whole network has introduced social funds in many ways, cooperated with public welfare organizations and local communities, and successively established community-based protection sites in the protection vacancy areas in Yunnan, supported local villagers to set up patrol teams, carried out field patrol monitoring, eliminated human interference and recorded wildlife traces and images.

Carry out habitat restoration and corridor construction in a scientific and orderly manner.

The habitat and corridor planning of Yunnan snub-nosed monkey was launched, some priority restoration areas were identified, social welfare funds were introduced, local tree species were used, forest vegetation on degraded patches was restored through scientific planning and management, and the habitats of five Yunnan snub-nosed monkey populations with serious fragmentation in the southern area were repaired and connected.

Explore the Yunnan snub-nosed monkey friendly community development project.

Demonstration villages were set up as friendly communities and protection-themed brand "MISIFI 弥司子" was created. A series of related activities were carried out so that the community can benefit from protection and promote people's real recognition, support and participation in Yunnan snub-nosed monkey and habitat protection.

Promote broad public participation.

The range-wide conservation network has established the protection brand of "Yunnan snub-nosed monkey 3000 plus" and held various forms of brand communication, public participation and fund-raising activities such as "selection of excellent Yunnan snub-nosed monkey front-line protection workers", online classes, theme photography exhibition, Yunnan snub-nosed monkey peripheral product development and charity sale, and webcast, so as to make more people understand the importance of Yunnan snub-nosed monkey and its protection, and attract potential supporters to establish a sustainable protection mechanism.

Excerpts

Kunming—Southwest China's Yunnan province has achieved remarkable results in the conservation of golden hair monkeys as the population of the species increased to more than 3,300 covering 23 varieties.

This was revealed by a green book on the outcomes of conservation efforts involving the endangered species in the province. It documents the entire process of comprehensive and systematic monitoring and evaluation of Yunnan golden hair monkeys, according to the provincial forestry and grassland bureau.

The Yunnan golden hair monkeys are listed as national first-class rare and endangered protected species in China and as vulnerable on the International Union for Conservation of Nature (IUCN) Red List. The monkeys live in the mountainous forests of the province and the neighboring Tibet Autonomous Region.

There were around 1,000 to 1,500 Yunnan golden hair monkeys of 13 varieties in 1996, and 3,000 individuals of 18 varieties in 2006, said the bureau.

"The protection of Yunnan golden monkeys has accumulated experience and set a good example for the conservation of flagship species," said Xiang Ruwu with the bureau.

(Excerpted from "Yunnan Snub-nosed Range-wide Conservation" from *China Daily*)

Vocabulary

Yunnan snub-nosed monkey 滇金丝猴
habitat 栖息地
network 网络
friendly community 友好社区

Reflection and Activities

Watch a video clip from the documentary "Green Plus 4: The Yunnan Snub-nosed monkey." Then try to finish the following exercises.

Source: https://www.xinpianchang.com/a10751719

1. After watching the video, try to summarize the timeline of the changes in people's behaviors and attitudes, and then try to draw a mind map about it.

2. Develop your mind map into a short presentation themed on "positive change through action", and share your understanding with your teammates.

第五节 滇金丝猴全境保护

导读

云南金丝猴是我国特有的濒危物种。它只能在西藏和云南交界处澜沧江和金沙江之间 7000 平方公里的高海拔森林中找到。栖息地所在的三江并流世界自然遗产地拥有 20% 的中国高等植物和 25% 的脊椎动物。它是全球公认的生物多样性模型之一。

云南金丝猴及其栖息地的全球保护有助于维护区域原始森林生态系统的完整性和丰富的生物多样性，对构建中国西南地区的生态安全屏障具有重要作用。

1. 挑战

40 年来，在政府和社会力量的共同努力下，云南金丝猴的保护工作取得了很大进展。金丝猴数量已从 20 世纪 90 年代第一次调查时的 1500 多增加到目前的 3000 以上。

然而，考虑到整个分布区的景观规模，它仍然面临许多挑战。对生物多样性保护和管理认识不足，如缺乏对生物多样性的统一监测；部分重点栖息地未纳入自然保护区体系，缺乏有效管理和保护；道路、村庄和农田导致栖息地碎片化，阻碍猴子之间的基因交换；当地居民高度依赖林下种植、采集、放牧等生产方式，当地经济发展与生态保护存在矛盾；保护投资与需求不匹配，公众意识和参与度低。

2. 联合保护机制

自 2019 年以来，云南省林业和草原局、云南省绿色环境发展基金会和自然保护协会等 13 家机构联合发起了"云南金丝猴大范围保护网络"，试图建立多方联合保护机制，并从巡查监测、走廊恢复、森林保护等方面采取行动，社区参与、友好发展和公众参与，为实现云南金丝猴种群的持续健康增长，栖息地生态系统健康水平全面提高。

生物多样性巡逻监测的标准化和信息化。

在巡更监控中，充分发挥了多个标准化监控方案、巡更监控在线数据管理平台和移动应用程序的作用，实现了巡更监控数据的无纸化、信息化、自动化采集、存储和分析。

建立社区保护区，以填补保护空缺。

整个网络通过多种方式引入社会资金，与公益组织和当地社区合作，先后在云南省保护空置区建立了社区保护点，支持当地村民成立巡逻队，进行现场巡逻监测，消除人为干扰，记录野生动物迹迹和图像。

科学有序地开展栖息地恢复和廊道建设。

启动云南金丝猴栖息地和走廊规划，确定一些优先恢复区域，引入社会福利基金，使用当地树种，通过科学规划和管理恢复退化斑块上的森林植被，对云南南部5个破碎化严重的金丝猴种群的栖息地进行了修复和连接。

探索云南金丝猴友好社区发展项目。

云南建立了友好社区示范村，创建了保护主题品牌"MISIFI弥司子"。开展了一系列相关活动，让社区受益于保护，促进人们对云南金丝猴和栖息地保护的真正认识、支持和参与。

促进公众广泛参与。

广域保护网络建立了"滇金丝猴3000+"保护品牌，并举办了"云南金丝猴优秀一线保护工作者评选"、网络课程、主题摄影展、文化节等多种形式的品牌传播、公众参与和募捐活动，云南金丝猴周边产品开发、慈善义卖、网络直播，让更多人了解云南金丝猴及其保护的重要性，吸引潜在支持者建立可持续保护机制。

节选

昆明——中国西南部的云南省，在保护金丝猴方面取得了显著成果，金丝猴的数量增加到3300多只，涵盖23个品种。

一本关于该省濒危物种保护工作成果的绿皮书揭示了这一点。据省林业和草原局称，该文件记录了对云南金丝猴进行全面、系统监测和评估的整个过程。

云南金丝猴被列为中国国家一级珍稀濒危保护物种，并被国际自然保护联盟（IUCN）列为濒危物种。猴子生活在该省和邻近的西藏自治区的山区森林中。

该局称，1996年云南金丝猴约有1000至1500只，分13个品种，2006年有3000只，分18个品种。

"云南金丝猴的保护积累了经验，为保护旗舰物种树立了良好的榜样，"该局的向如武说。

（节选自《中国日报》报道：云南开展金丝猴保护）

思考题

观看以下这段关于保护云南金丝猴的视频，然后完成以下练习。

视频来源：https://www.xinpianchang.com/a10751719

1. 看完视频后，总结人们行为和态度变化的时间线，然后画一张思维导图。

2. 将你的思维导图拓展成一个以"行动带来积极变化"为主题的简短演讲，并与小组成员分享你的理解。

Section VI: Ecological Restoration of Qingxi Country Park in Shanghai

Introduction

In the process of rapid urbanization development, the urban functions of metropolises have been continuously improved, the scale of cities has been expanding, the demand for ecological recreation of citizens has risen rapidly, and the contradictions of tight constraints on natural resources, insufficient supply of ecological space and relatively low quality have been further highlighted.

Since the 1990s, with the continuous acceleration of the urbanization process, human production, life and other construction activities have increased sharply, and multiple problems such as disorderly land layout, shrinking wetlands, river siltation and agricultural non-point source pollution have appeared in the Dianshan Lake area.

Before the completion of Qingxi Country Park, the lake body of Dalian Lake within the existing planning range shrank, the wetland area decreased, and the biodiversity of fish, benthic animals and other biodiversity decreased significantly; the artificial division of aquaculture in the lake area, the siltation of the river, the serious decline in water connectivity, coupled with the discharge of sewage from industry, domestic and other sewage and farmland fertilization and irrigation, the phenomenon of non-point source pollution was serious; the layout of industrial land in the region was scattered, the use of extensive and inefficient, and the style of Jiangnan water town was gradually lost.

In 2012, Shanghai launched the planning and site selection of country parks. Qingxi Country Park is located in the Dianshan Lake area on the border of Shanghai with Jiangsu and Zhejiang, in the southwest of Qingpu District, Shanghai, Jinze town and Zhujiajiao town, with a planned total area of about 21.85 square kilometers. The area is home to 21 natural lakes in Shanghai, also an important water source protection area and ecological protection area in Shanghai. The country park is rich in species resources, "lakes, banks, lakes, islands" crisscrossed. It is the only country park in Shanghai featuring wetlands as its diverse aquatic species can be called Shanghai's natural native aquatic species gene bank. The park completed the first phase of construction and opened to the public in October 2016, with an open area of 4.65 square kilometers, with the theme of wetland, ecology, nature and leisure, of which more than 60 acres of water forest is unique to Shanghai, known as the wonder of *Taxodium ascendens*.

Excerpts

In the construction of Shanghai Qingxi Country Park, two major measures were mainly taken to "treat" Dianshan Lake.

The first is to adhere to the guidance of planning and fully implement the concept of ecological priority. The implementation of the project attaches great importance to the leading role of rural unit village planning. The design concept fully absorbs the results of the international program collection and highlights the concept of ecological priority and systematic restoration, in accordance with the specific principles of "respecting nature, giving priority to protection, scientific restoration, appropriate development, and rational utilization", while adhering to the overall planning of all elements of water, forest, field, lake, grass, village, and factory, planning and clearing functional divisions, and maintaining the existing river and lake water system, farmland forest network, natural villages and other textures as the characteristics. The design concept highlights the protection and restoration of water, forests and fields, combining the shaping of regional spatial and cultural characteristics, trying to build up a suburban wetland type country park with the main functions of ecological conservation, wetland science popularization, agricultural production, experience and leisure.

The second is to implement various ecological protection and restoration projects. On the basis of implementing the requirements of planning functional zoning, the Qingxi Country Park project mainly adopts measures such as wetland protection and natural restoration, adjustment of land use structure layout, farmland ecosystem improvement, comprehensive improvement of river channels, and shaping of popular science leisure and cultural space to restore the stability of regional ecosystems and improve the overall ecological quality of the region.

The construction of Qingxi Country Park has shown obvious results in three aspects.

The first is to optimize the layout of the land-use structure. Through the demolition and reclamation of inefficient construction land and the rectification of pit pond water surface and aquaculture water surface, while completely removing pollution sources and alleviating non-point source pollution, about 0.8 hectares of new forest land and about 99.2 hectares of new cultivated land have been added, and the regional land-use structure has been continuously optimized.

Second, the core function of regional ecological green is prominent. Through the systematic restoration and comprehensive treatment of landscapes, forests and lakes, the green ecological space based on water, forests and fields has been anchored, the new water area is about 19.8 hectares, and the ecosystem structure of wetlands and farmlands has been continuously improved. What's more the connectivity of the water system has been enhanced, the biodiversity has been maintained while the overall ecological environment and ecological function of the region have been significantly improved, and the citizens have also provided a natural and wild leisure space.

The third is to achieve the transformation of ecological advantages. Regional ecological advantages have gradually been transformed into development advantages, and Huawei's mobile terminal R&D center has been settled, which becomes an important engine of promoting regional leapfrog development. Thus a green development path based on high-end R&D, eco-tourism, and modern agriculture has gradually taken shape.

(Excerpted from Colomn of "Nature's Typical Cases of Ecological Restoration in China")

Vocabulary

country park 郊野公园

systematic restoration and comprehensive treatment 系统修复和综合治理

promote regional leapfrog development 促进区域跨越式发展

R&D center: research and development center 研究与发展中心

Reflection and Activities

1. Work in groups of 4-6 students. Make a public service announcement video about protecting the ecological balance within 5 minutes and design one public service slogan for the theme of your group's video.

2. Appreciate and read an ancient poem with traditional Chinese culture that describes the harmonious coexistence between man and nature. Make class presentation of your poetry reading with proficiency in skilled recitation, excellence in voice and feelings, and in line with the rhythm of poetry.

Return to Nature

Tao Yuanming

While young, I was not used to worldly cares,

And hills became my natural compeers.

But by mistake I fell in mundane snares,

And was thus entangled for thirteen years.

A caged bird would long for wonted wood,

And fish in ponds for native pools would yearn.

Go back to till my southern field I would,

To live a rural life why not return?

My plot of ground is but ten acres square;

My thatched cottage has eight or nine rooms.

In front I have peach trees ere and plums there;

Over back eaves willows and elms cast glooms.

A village can be seen in distant dark,

Where plumes of smoke rise and waft in the breeze.

In alley deep a dog is heard to bark,

And cocks crow as if over mulberry trees.

Into my courtyard no one should intrude,

Nor rob my private rooms of peace and leisure.

After long, long official servitude,

Again in nature I find homely pleasure.

(Translated by Xu Yuanchong)

第六节 上海青西郊野公园生态修复

导读

在快速城镇化发展过程中，大都市的城市功能不断提升，城市规模不断扩大，市民的生态游憩需求快速上涨，自然资源紧约束、生态空间供给不足且品质相对不高等矛盾进一步凸显。

20世纪90年代以来，随着城镇化进程的不断加快，人类生产、生活等建设活动骤增，淀山湖区域出现用地布局无序、湿地萎缩、河道淤积及农业面源污染等多重问题。

青西郊野公园建成前，现有规划范围内大莲湖湖体萎缩，湿地面积减少，鱼类、底栖动物等生物多样性明显下降；湖区人为分割养殖，河道淤积，水体连通性严重下降，加之工业、生活等污水排放及农田施肥灌排，面源污染现象严重；区域内工业用地布局散乱，利用粗放、低效，江南水乡风貌渐失。

2012年，上海市启动了郊野公园规划选址工作。青西郊野公园位于上海与江苏、浙江交界的淀山湖地区，在上海市青浦区西南部的金泽镇和朱家角镇境内，规划总面积约21.85平方公里。该地区聚集着上海市21个自然湖泊，是上海重要的水源保护地和生态保护区，郊野公园内物种资源丰富，"湖、滩、荡、岛"纵横交错，是上海市唯一一个以湿地为特色的郊野公园，其多样的水生生物物种堪称上海天然的本土水生物种基因库。公园于2016年10月完成一期建设并对外开放，开放区4.65平方公里，以湿地、生态、自然、休憩为主题，其中60多亩的水上森林上海独有，被誉为池杉奇观。

节选

上海青西郊野公园建设过程中，"救治"淀山湖主要采取了两大主要措施。

一是坚持规划引领，全面贯彻生态优先理念。项目实施高度重视郊野单元村庄规划引领作用。设计理念充分吸收国际方案征集成果，突出生态优先、系统修复理念，按照"尊重自然、保护优先、科学修复、适度开发、合理利用"的具体原则，坚持水、林、田、湖、草、村然村落等肌理为特色，突出水、林、田为主的保护修复，结合地区空间人文特色塑造，打造以生态保育、湿地科普、农业生产、体验休闲为主要功能的远郊湿地型郊野公园。

二是实施各类生态保护修复工程。在落实规划功能分区要求基础上，青西郊野公园项目主要采取湿地保护与自然恢复、用地结构布局调整、农田生态系统整治、河道综合整治、科普休闲人文空间塑造等措施，恢复区域生态系统稳定性，提升区域整体生态品质。

青西郊野公园的建设，在三个方面呈现出明显成效。

一是用地结构布局优化。通过低效建设用地拆除复垦以及坑塘水面、养殖水面等整治，彻底清除污染源、缓解面源污染的同时，新增林地约 0.8 公顷，新增耕地约 99.2 公顷，区域用地结构不断优化。

二是区域生态绿核功能凸显。通过山水林田湖草的系统修复和综合治理，锚固了以水、林、田为基底的绿色生态空间，新增水域面积约 19.8 公顷，湿地、农田生态系统结构不断改善，水系连通性增强，生物多样性得到维护，区域整体生态环境和生态功能显著提升，也为市民提供了自然、野趣的休闲游憩空间。

三是生态优势实现转化。区域生态优势逐步转化为发展优势，华为移动终端研发中心落户，成为推动区域跨越式发展的重要引擎，一条以高端研发、生态旅游、现代农业为基础的绿色发展路径逐步形成。

（节选自学习强国《中国生态修复典型案例》专栏：上海青西郊野公园生态修复）

思考题

1. 4 到 6 人一组制作 5 分钟以内的关于保护生态平衡的公益广告视频，并为本组视频的主题设计一条公益广告语。

2. 欣赏、朗诵中国传统文化中描述人与自然和谐相处的古诗词《归园田居》（其一）。班级展示，要求朗诵熟练，声情并茂，配乐贴合诗歌韵律。

归园田居·其一

陶渊明

少无适俗韵，性本爱丘山。
误落尘网中，一去三十年。
羁鸟恋旧林，池鱼思故渊。
开荒南野际，守拙归园田。
方宅十余亩，草屋八九间。
榆柳荫后檐，桃李罗堂前。

暧暧远人村，依依墟里烟。
狗吠深巷中，鸡鸣桑树颠。
户庭无尘杂，虚室有余闲。
久在樊笼里，复得返自然。

（许渊冲译）

Section vii: Carbon Peak and Carbon Neutrality

Introduction

In 1992, China became one of the first parties to the United Nations Framework Convention on Climate Change (hereinafter referred to as the Convention). In dealing with climate change, China adheres to the principles of common but differentiated responsibilities, fairness and respective capabilities, and resolutely safeguards the rights of developing countries, including China. In 2002, the Chinese government approved the Kyoto Protocol. In 2007, the Chinese government formulated the "China's national programme for coping with climate change", which defined the specific objectives, basic principles, key areas and policy measures for coping with climate change by 2010, and required that the energy consumption per unit of GDP in 2010 should be reduced by 20% compared with that in 2005. In 2007, the Ministry of Science and Technology, the National Development and Reform Commission and other 14 departments jointly formulated and released the "China's special action on science and technology to cope with climate change", proposing the goals, key tasks and safeguard measures for the development of science and technology and the improvement of independent innovation capacity in the field of coping with climate change by 2020.

In November, 2013, China issued the first strategic plan for adapting to climate change, the National Strategy for Climate Change Adaptation, which made the systems and policies for coping with climate change more systematic. In June, 2015, China submitted to the Convention secretariat the document "Enhanced Actions on Climate Change: China's Intended Nationally Determined Contributions", which set the goal of independent action by 2030: carbon dioxide emissions will peak around 2030 and strive to reach the peak as soon as possible; carbon dioxide emissions per unit of GDP decreased by 60%—65% compared with 2005, non-fossil energy accounted for about 20% of primary energy consumption, and forest reserves increased by about 4.5 billion cubic meters compared with 2005. With China's active promotion, countries around the world reached the Paris Agreement on climate change in 2015. China has won the best interests for developing countries in terms of independent contribution, financing, technical support, transparency and so on. In 2016, China took the lead in signing the Paris Agreement and actively promoted its implementation. By the end of 2019, China has exceeded the 2020 climate action goal ahead of schedule, and has established the image of a big country that keeps its commitments.

Excerpts

In September 2020, President Xi Jinping stated at the general debate of the seventy-fifth session of the United Nations General Assembly that, in response to climate change, the Paris Agreement represents the general chirection of a global green and how-carbon transition, and that is the minimum action needed to protect the Earth. All countries must take decisive steps forward.

At the same time, it was announced that China would improve its national independent contribution, adopt more powerful policies and measures, strive to peak carbon dioxide emissions by 2030, and strive to achieve carbon neutrality by 2060.

At the Climate Ambition Summit held in December 2020, President Xi Jinping further announced that by 2030, China will its carbon dioxide emissions per unit of GDP by over 65% from the 2005 level, increase the shore of non-fossil fuels in primary energy consumption to around 25%, increase the forest stock volume by 6 billion cubic meters from the 2005 level, and bring its total installed capacity of wind and solar power to over 1.2 billion kilowatts.

To achieve the "double carbon" goal, we must do it ourselves rather than others.

The Paris Agreement on climate change represents the general direction of global green and low-carbon transformation, and is the minimum action that needs to be taken to protect the earth's home. Countries must take decisive steps. China will increase its national independent contribution, adopt more effective policies and measures, strive to peak carbon dioxide emissions by 2030, and strive to achieve carbon neutrality by 2060.

——Xi Jinping's speech at the general debate of seventy-fifth session of the UNGA, September 22th 2020

I have announced that China will strive to peak carbon dioxide emissions by 2030 and achieve carbon neutrality by 2060. To achieve this goal, China needs to make extremely arduous efforts. We believe that as long as it is something beneficial to all mankind, China should be duty bound to do it well. China is formulating an action plan and has begun to take concrete measures to ensure the achievement of the set goals. In doing so, China is practicing multilateralism with practical actions and contributing to the protection of our common homeland and the realization of sustainable human development.

——Xi Jinping's speech at the World Economic Forum's "Oavos Agenda" dialogue, January 25th 2021

Promoting carbon peak and carbon neutralization is a major strategic decision made by the Central Party Committee after careful consideration. It is not only our solemn commitment to the international community, but also the internal requirement of promoting high-quality development.

——Xi Jinping's speech at Central Economic Work Conference, December 8th 2021

To achieve the "double carbon" goal, we must do it ourselves rather than others. China has entered a new stage of development. Promoting the "double carbon" work is an urgent need to solve the prominent problems of resource and environmental constraints and achieve sustainable development. It is an urgent need to comply with the trend of technological progress and promote the transformation and upgrading of economic structure. It is an urgent need to meet the growing needs of the people for a beautiful ecological environment and promote the harmonious coexistence between man and nature. It is an urgent need to take the initiative to assume the responsibility of a big country. There is an urgent need to promote the construction of a community with a shared future for mankind.

——Xi Jinping's speech at the 36th study session of Political Bureau of the CPC Central Committee, January 24th, 2022

Vocabulary

carbon peak 碳达峰

carbon neutrality 碳中和

carbon trade 碳交易

carbon sink 碳汇

Reflection and Activities

1. Why does China make a double carbon commitment?
2. How should our school build a carbon neutral campus?

第七节 7　碳达峰和碳中和

导读

1992年，中国成为最早签署《联合国气候变化框架公约》（以下简称公约）的缔约方之一。在应对气候变化问题上，中国坚持共同但有区别的责任原则、公平原则和各自能力原则，坚决捍卫包括中国在内的广大发展中国家的权利。2002年中国政府核准了《京都议定书》。2007年中国政府制定了《中国应对气候变化国家方案》，明确到2010年应对气候变化的具体目标、基本原则、重点领域及政策措施，要求2010年单位GDP能耗比2005年下降20%。2007年，科技部、国家发展改革委等14个部门共同制定和发布了《中国应对气候变化科技专项行动》，提出到2020年应对气候变化领域科技发展和自主创新能力提升的目标、重点任务和保障措施。

2013年11月，中国发布第一部专门针对适应气候变化的战略规划《国家适应气候变化战略》，使应对气候变化的各项制度、政策更加系统化。2015年6月，中国向公约秘书处提交了《强化应对气候变化行动——中国国家自主贡献》文件，确定了到2030年的自主行动目标：二氧化碳排放2030年左右达到峰值并争取尽早达峰；单位国内生产总值二氧化碳排放比2005年下降60%—65%，非化石能源占一次能源消费比重达到20%左右，森林蓄积量比2005年增加45亿立方米左右。在中国的积极推动下，世界各国在2015年达成了应对气候变化的《巴黎协定》，中国在自主贡献、资金筹措、技术支持、透明度等方面为发展中国家争取了最大利益。2016年，中国率先签署《巴黎协定》并积极推动落实。到2019年底，中国提前超额完成2020年气候行动目标，树立了信守承诺的大国形象。

节选

2020年9月，习近平主席在第七十五届联合国大会一般性辩论上阐明，应对气候变化，《巴黎协定》代表了全球绿色低碳转型的大方向，是保护地球家园需要采取的最低限度行动，各国必须迈出决定性步伐。

同时宣布，中国将提高国家自主贡献力度，采取更加有力的政策和措施，二氧化碳排放力争于2030年前达到峰值，努力争取2060年前实现碳中和。

在 2020 年 12 月举行的气候雄心峰会上，习近平主席进一步宣布，到 2030 年，中国单位国内生产总值二氧化碳排放将比 2005 年下降 65% 以上，非化石能源占一次能源消费比重将达到 25% 左右，森林蓄积量将比 2005 年增加 60 亿立方米，风电、太阳能发电总装机容量将达到 12 亿千瓦以上。

实现"双碳"目标，不是别人让我们做，而是我们自己必须要做。

应对气候变化《巴黎协定》代表了全球绿色低碳转型的大方向，是保护地球家园需要采取的最低限度行动，各国必须迈出决定性步伐。中国将提高国家自主贡献力度，采取更加有力的政策和措施，二氧化碳排放力争于 2030 年前达到峰值，努力争取 2060 年前实现碳中和。

——2020 年 9 月 22 日，习近平在第七十五届联合国大会一般性辩论上的讲话

我已经宣布，中国力争于 2030 年前二氧化碳排放达到峰值、2060 年前实现碳中和。实现这个目标，中国需要付出极其艰巨的努力。我们认为，只要是对全人类有益的事情，中国就应该义不容辞地做，并且做好。中国正在制定行动方案并已开始采取具体措施，确保实现既定目标。中国这么做，是在用实际行动践行多边主义，为保护我们的共同家园、实现人类可持续发展作出贡献。

——2021 年 1 月 25 日，习近平在世界经济论坛"达沃斯议程"对话会上的特别致辞

推进碳达峰碳中和是党中央经过深思熟虑作出的重大战略决策，是我们对国际社会的庄严承诺，也是推动高质量发展的内在要求。

——2021 年 12 月 8 日，习近平在中央经济工作会议上的讲话

实现"双碳"目标，不是别人让我们做，而是我们自己必须要做。我国已进入新发展阶段，推进"双碳"工作是破解资源环境约束突出问题、实现可持续发展的迫切需要，是顺应技术进步趋势、推动经济结构转型升级的迫切需要，是满足人民群众日益增长的优美生态环境需求、促进人与自然和谐共生的迫切需要，是主动担当大国责任、推动构建人类命运共同体的迫切需要。

——2022 年 1 月 24 日，习近平在十九届中央政治局第三十六次集体学习时的讲话

思考题

1. 中国为何作出双碳承诺？
2. 我校应该如何打造碳中和校园？

Section viii: China Embraces Garbage Classification, with Shanghai Taking the Lead

Garbage sorting illustration in Shanghai

Introduction

From July 1st, 2019, Shanghai residents are required by law to sort garbage into four different categories, or they could face fines. Shanghai's most heated topic of "which is the right bin for my garbage" may sweep other cities in the near future, as the national legislation is planning to speed up mandatory garbage sorting by writing it into law. Shanghai is among China's first cities to introduce garbage classification and also the country's most serious in its implementation. Its regulation covers reducing the amount of garbage produced at sources, ensuring separate transportation of different trashes, upgrading treatment and promoting social participation.

1. The embodiment of Xi Jinping's ecological civilization thoughts

Since the 18th Communist Party of China (CPC) National Congress, the CPC Central Committee with General Secretary Xi Jinping at its core has been advancing ecological progress in an all-round way, bringing about historic, transformable and overarching changes in environmental protection. At the 19th CPC National Congress, Xi Jinping Thought on Ecological civilization was put forward as a fundamental plan for the sustainable development of the Chinese nation, and pointed out that the ecological environment is a major political issue that bears on the mission and purpose of the CPC.

At present, the construction of ecological civilization in China is in the critical period. As one of the main focuses of ecological civilization construction, garbage classification is particularly important. In December 2016, General Secretary Xi Jinping made important instructions at the meeting of the Central Leading Group for Financial and Economic Affairs to universally implement the garbage classification system, and required Beijing, Shanghai and other cities to catch up with the international level, take the lead in establishing a mandatory garbage classification system, to set an example for the whole country. In November 2018, Xi Jinping, General Secretary of the CPC, stressed during an inspection tour of Jiaxing Road, Shanghai, that garbage sorting is a new fashion, and he hoped that Shanghai would work hard to implement it. In October 2019, the communique of the fourth Plenary Session of the 19th Communist Party of China explicitly proposed the universal implementation of garbage sorting and resource utilization system. Garbage classification is the general trend, which should also become an important task of health promotion.

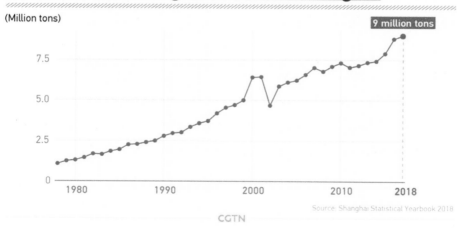

Shanghai Statistical Yearbook 2018

2. Shanghai's garbage problem

With the development of China's social economy, people's material life is greatly enriched, and the amount of domestic garbage is increasing year by year. According to the 2018 Annual Report on the Prevention and Control of Environmental Pollution by Solid Waste in Large and medium-sized Cities published by the Ministry of Ecology and Environment in December 2018, the amount of domestic waste produced in 202 large and medium-sized cities in 2017 was 20.944 million tons, an increase of 7.1% over 2016. According to the Statistical Yearbook of Shanghai Greening City Appearance, in 2017, Shanghai produced 9.843 million tons of household garbage, an increase of 9.4% compared with 2016. With about 27,000 tons of waste, it takes two weeks to build a "Jin Mao Tower". If the huge amount of garbage can not be orderly disposed of, the problem of "garbage siege" will become more prominent, and the beautiful home will be surrounded by garbage.

Shanghai began a pilot program of garbage sorting in Caoyangwu Village, Putuo District in 1995. In 1999, the Shanghai Municipal Appearance and Environmental Sanitation Administration formulated the Implementation Plan for the Classification, Collection and Disposal of Domestic Garbage in Shanghai Urban Area, which determined the standards for the classification of domestic garbage into organic, inorganic, toxic and harmful wastes. In June 2000, the Ministry of Construction issued a "notice on the announcement of the pilot city of household garbage classification collection", which was officially the prelude of the pilot work of garbage classification collection in China. Since 2011, Shanghai began to promote a new round of waste classification and reduction work. This work was promoted as a practical project of Shanghai Municipal government in 2007. In February 2014, Shanghai Municipality issued the "Shanghai Municipal Measures for Promoting the Classification and Reduction of Household Garbage" in the form of government regulations. The four classification standards of hazardous waste, wet waste and dry waste were solidified, which also provided guidance for the introduction of a series of garbage classification systems later. In January 2019, "Regulations of Shanghai Municipality on the Management of Household Garbage" were approved by the second session of the 15th Shanghai Municipal People's Congress, marking stage of the comprehensive implementation of garbage classification and governance by law in Shanghai. From individual pilot project to universal implementation in the whole city, it has been 24 years since Shanghai embarked on a path of garbage classification.

Excerpts

As early as in 2000, Beijing, Shanghai and some other cities became the first pilot cities to advocate garbage sorting. However, 17 years later, the effects of garbage sorting in China are far from satisfactory.

China's National Development and Reform Commission and Ministry of Housing and Urban-Rural Development issued "The Implementation Plan of Garbage Sorting" in March, which requires 46 cities nationwide to carry out mandatory garbage sorting, and guides residents to sort garbage on their own. The plan also proposes to establish a basic system of laws and regulations on waste classification by the end of 2020.

The garbage disposal capacity fails to catch up with the speeding up of garbage producing in China. Research shows that there would be no place in Beijing to bury garbage in four to five years. In Shanghai, some landfill sites have even been built close to residential areas.

Mao Da, one of the founders of the volunteering activity "zero waste allisnce", suggested the government inform the public of the cost of not sorting garbage. He said in an interview that if the public know how they might be hurt by not doing this, they would feel more pressure and motivation to think it, as residents are also struggling with how to sort waste properly. For example, the right way to sort a cup of leftover bubble tea and half-eaten crayfish has sparked great debate on social media.

Other problems also include harsh time and place requirements for trash dumping and improper disposal of sorted garbage, as some cases in Shanghai showed. But the situation is improving. The municipal government has set up online apps to handle sorting inquiries, and issued guidelines to address the "one-size-fits-all" method. Cities like Beijing and Shanghai have launched point-rewarding scheme to motivate its residents by allowing them to exchange for daily products with points earned through proper sorting.

(Excerpted from "China enforces garbage sorting in 46 cities"—CGTN)

Vocabulary

pilot city 试点城市
mandatory 义务的；强制的
dump 丢弃

urban-rural 城乡
zero waste 零废弃

Reflection and Activities

1. Observe and classify your family's household garbage, then distinguish what kind of garbage are they.

2. Watch a video on the introduction tour of the garbage sorting showroom in Hebei University of Environmental Engineering and draw a flowchart of the necessary steps of garbage disposal.

第八节 中国采用垃圾分类——上海试点

导读

从 2019 年 7 月 1 日开始，上海市民被法律要求将垃圾分为四类，否则他们可能面临罚款。上海最热门的话题"我的垃圾属于哪种垃圾"可能在不久的将来席卷其他城市，因为国家立法机构正计划通过将其写入法律来加快强制垃圾分类。上海是中国首批引入垃圾分类的城市之一，也是中国实施垃圾分类最为严格的城市之一。其规定包括从源头上减少垃圾产生量，确保不同类别垃圾的单独运输，升级处理和促进社会参与。

1. 习近平生态文明思想的体现

党的十八大以来，以总书记为核心的党中央全面推进生态文明建设，实现了环境保护的历史性、变革性、全局性变革。党的十九大提出生态文明思想是中华民族可持续发展的根本大计，指出生态环境是关系党的使命和宗旨的重大政治问题。

当前，我国生态文明建设正处于关键时期。垃圾分类作为生态文明建设的重点之一，显得尤为重要。2016 年 12 月，总书记在中央财经领导小组会议上作出重要指示，全面推行垃圾分类制度，要求北京、上海等城市赶超国际水平，率先建立强制性垃圾分类制度，为全国树立榜样。2018 年 11 月，总书记在上海嘉兴路视察时强调，垃圾分类是一种新的时尚，希望上海努力落实。2019 年 10 月，中共十九届四中全会公报明确提出，普遍推行垃圾分类和资源利用制度。垃圾分类是大势所趋，也应成为健康促进的重要任务。

2. 上海的垃圾困境

随着我国社会经济的发展，人们的物质生活大大丰富，生活垃圾量也在逐年增加。根据 2018 年 12 月生态环境部发布的《2018 年大中城市固体废物污染环境防治年报》，2017 年 202 个大中城市生活垃圾产生量为 2094.4 万吨，比 2016 年增长 7.1%。根据《上海市绿化市容统计年鉴》，2017 年，上海市产生生活垃圾 984.3 万吨，比 2016 年增长 9.4%。大约有 27000 吨垃圾，堆成一座"金茂大厦"需要两周的时间。如果大量垃圾不能得到有序处置，"垃圾围城"问题将更加突出，美丽的家园将被垃圾包围。

1995 年，上海开始在普陀区曹杨吴村进行垃圾分类试点。1999 年，上海市容环卫局制定了《上海市区生活垃圾分类、收集和处置实施方案》，确定了生活垃圾分类为有机、无机、有毒和有害废弃物的标准。2000年6月，建设部发布"关于公布生活垃圾分类收集试点城市的通知"，正式拉开了我国垃圾分类收集试点工作的序幕。2011年以来，上海开始推进新一轮垃圾分类减量工作。这项工作于2007年被上海市政府作为一项实际项目推广。2014年2月，上海市以政府规章的形式发布了《上海市推进生活垃圾分类减量办法》。固化了危险废物、湿废物和干废物的四个分类标准，这也为以后引入一系列垃圾分类系统提供了指导。2019年1月，《上海市生活垃圾管理条例》经上海市第十五届人民代表大会第二次会议批准通过，标志着上海市进入全面实施垃圾分类、依法治理阶段。从个别小区试点到全市普遍实施，上海走上垃圾分类道路已经24年了。

节选

早在2000年，北京、上海等城市就成为首批倡导垃圾分类的试点城市。然而，17年过去了，我国垃圾分类的效果却远远不能令人满意。

国家发展改革委和住房城乡建设部于2017年3月发布了《生活垃圾分类制度实施方案》，要求全国46个城市实行垃圾强制分类，并引导居民自行分类。该计划还提议到2020年底建立一个关于垃圾分类的基本法律法规体系。

我国垃圾处理能力跟不上垃圾生产速度的加快。有研究显示，在四到五年内，北京将没有地方掩埋垃圾。在上海，一些垃圾填埋场甚至建在靠近居民区的地方。

"零废弃"志愿活动的创始人之一毛达建议政府告知公众不分类垃圾的费用。他在一次采访中说，如果公众知道不这样做可能会对他们造成多大的伤害，他们会更有压力和动力，因为居民们也在努力解决如何正确分类垃圾的问题。例如，如何正确分类一杯剩下的珍珠奶茶和吃了一半的小龙虾在社交媒体上引发了激烈的争论。

其他问题还包括垃圾倾倒的严格时间和地点要求以及分类垃圾的不当处理，如上海的一些案例所示。但情况正在改善。市政府已经建立了在线应用程序来处理分类查询,并发布了解决"一刀切"方法的指导方针。北京和上海等城市已经推出了积分奖励计划，以激励其居民，允许他们用通过适当分类获得的积分兑换日常产品。

（节选自中国国际电视台报道：中国在46个城市实施垃圾分类）

思考题

1. 观察和分类你家的家庭垃圾，区分它们是什么类型的垃圾。
2. 观看河北环境工程学院垃圾分类展厅介绍垃圾分类的视频,绘制垃圾处理的必要步骤流程图。